Virusphere

Virusphere

From Common Colds to Ebola Epidemics

Frank Ryan

 Prometheus Books

Guilford, Connecticut

 Prometheus Books

An imprint of Rowman & Littlefield Publishing Group, Inc.
4501 Forbes Blvd., Ste. 200
Lanham, MD 20706
www.rowman.com

Library of Congress Cataloging-in-Publication Data Is Available

ISBN 978-1-63388-604-9 (cloth: alk. paper)
ISBN 978-1-63388-605-6 (electronic)

∞™ The paper used in this publication meets the minimum requirements of American
National Standard for Information Sciences—Permanence of Paper for Printed
Library Materials, ANSI/NISO Z39.48-1992

I would like to thank my editors,
Myles Archibald and Hazel Eriksson,
and my agent, Jonathan Pegg,
for their support of my writing this book

We all play hideous games with each other.
We step inside each other's chalk circles.

Anthony Hopkins

Contents

Introduction

What springs to your mind when you sit back and consider viruses? I am altogether aware that even if you come, as will many of my readers, from a non-scientific background, you will likely be uncertain as to the strange and ill-definable nature of viruses. Even among scientists, viruses are among the most enigmatic of the biological entities that are to be found on our precious blue-girdled planet. Certainly, there is a good deal of misinformation about them. You might, for example, be apt to confuse them with bacteria, a confusion that is not helped by the fact that we doctors caused this confusion in the first case by lumping two radically different entities together as the 'microbes' that are the root cause of infectious diseases.

Viruses frighten us. They elicit a primal fear of the unknown. They are capable of crashing through our natural barriers and defences, turning healthy cells into microscopic factories to produce exponential numbers of daughter viruses. These swarm through the bloodstream, drawing the attention of the immune system. This provokes our own white blood cells and other immune defences to become an aggressive reactionary force, fixated on annihilating the invaders, regardless of the devastation in their wake – indeed our very own immune system contributes to many of the familiar symptoms of the resultant illness, from runny noses to violent, bloody haemorrhages. In effect, every infection becomes a pitched battle that will determine the outcome for us, the host. As we come to know viruses better, we discover that they have also come to know us intimately, taking advantage of our very behaviour to facilitate their infectiousness and spread among us. All in all, that is rather scary. Especially so when we consider that some of those viruses include some of the most dangerous entities on the planet, entities eminently capable of making us very sick indeed – even killing us.

It is hardly surprising that we should fear them and consider them an epitome of menace, perhaps even evil. But contrary to what we might think about viruses, they are not driven by malice. Their ultimate goal, just like that of all living organisms, is simply to survive and multiply, thereby ensuring the success of their kind. That's all very nice to know but this lack of malice is hardly a comfort to us when we are infected by a virus. It is only natural in such circumstances that we might resent viruses and what they might do to us. It is equally natural that we should also feel the need to protect those we love from viruses while knowing that, if and when one of them comes along, it will arrive among us, unseen and unknowable, causing damage and pain among us, even to the most innocent, without feeling or rational explanation.

We are comforted by the knowledge that there are vaccinations available to protect us and our families from these viruses. But that same misinformation about various different vaccination campaigns has sown confusion among us. Meanwhile, there appears to be little to nothing that we can do about certain epidemic varieties, such as the norovirus or the common cold, which sweep through our populations as common experiences, albeit experiences we would rather avoid, provoking altogether negative associations in relation to viruses. Is it any surprise that we are inclined to wonder what other capacity for mischief viruses are capable of? You might well take the question further and wonder aloud: what is the purpose of these minuscule entities? Wouldn't our world be a good deal better off if these dratted parasites were simply eliminated from the scene, never to bother us any more?

The curious thing is that, while as a doctor I sympathise with such sentiments, I cannot entirely agree with them. Counter-intuitive as it might seem, I know that a world without viruses would not be one in which I would care to live. Why do I suggest such a thing? It introduces what might appear to be a contradictory question. Does the existence of viruses really matter to you and to me? I can assure my readers that, contrary to what our primal instincts might suggest, the answer is, strangely, but indubitably, yes!

In his elegant essay 'The Lives of a Cell', the late Lewis Thomas drew attention to the fact that we humans are not above the rest of life. In his words, 'Man is embedded in nature.' While accepting the fundamental truth in this, I would further extrapolate from it the fact that nature is embedded in humankind. And the nature I am referring to includes viruses. I don't, however, expect you, my readers, to simply take my word on this. But even the possibility that we humans owe our existence to, and remain dependent on, the very existence of viruses is more than enough to suggest we take the trouble to understand them.

That is what I hope to achieve in writing this book. It is my intention to enable you, my readers, to understand viruses.

In fact, even at a self-interest level, it is only natural that we should feel the need to understand entities that are capable of seriously hurting us. We surely need to know how they might threaten us and our loved ones and what we can possibly do to mitigate this threat. I freely confess that this was my perspective when, as a youthful medical student, I performed my first experiments on viruses long ago. Years later it remained my perspective when, as a busy hospital physician dealing with serious medical emergencies, it was my duty to diagnose and treat many patients suffering from viral infections. But then, during a period of international travel aimed at examining what was actually happening when new 'emerging viruses' were threatening entire populations, I discovered that my former perspective was somewhat blinkered in relation to viruses. Here I invite my readers to remove those same blinkers and look beyond the undoubted mischief of viruses to find answers to their behaviour and presence among us.

Key to such understanding is to get to know their strange and extraordinary world. Thanks to the most modern advances in the science of virology, we are now in a better position than ever before to understand that world. We are about to explore what, in scientific circles, is now called 'the virosphere', a play on which forms the title of this book.

How could such an exploration be anything other than an odyssey? Viruses circumambulate our world with the speed of a passenger jet, paying no heed to national boundaries, or circumscribing notions of nationality, ethnicity, race or religion. They pay no heed to sex or age, or social class, or indeed any human hubris of fame, celebrity, wealth or power. They are devoid of any sense of justice or morality, so that our social or religiously derived morals prompting concepts of goodness, badness, justice, sin, simply do not apply. Now add to this situation the fact that these threatening entities are, for the most part, utterly invisible, even under the most powerful magnification of the light microscope, making them all the more enigmatic – and perhaps also the more scary. These invisible entities invade not just our tissues and organs, but, to borrow the metaphor from Anthony Hopkins, they step inside the chalk circles of our most intimate and innermost being: the nucleus of our living cells, the repository of our coding DNA.

It is not a bad thing that we should be wary of viruses, but in this exploration we need to rise above mere scare stories. The fact is that the vast majority of viruses in this world – and there really are very many of them – have no mischievous interest in humans. This rather begs a new question: if such viruses are not interested in you and me, what are they interested in?

My purpose in writing this book is to explain what viruses really are, to accurately define them, and from such an understanding to explore their role in our human history and the wider history of our world. I hope to do so in a way that readers, whether they come from a scientific background or have no prior knowledge of viruses, will understand the true importance of viruses to life and to the biosphere. Given such potential importance, how can we possibly get to grips with such minuscule entities? How can we even hope to visualise their quintessentially viral way of life in their complex ultramicroscopic world? As with historic explorations of alien worlds, it might help our exploration if we could avail ourselves of some guide.

I would suggest that there is an obvious guide, one that is entirely apposite, whose perspective has only recently been opened up to us in wonderful detail through the deepest penetration of the living world by the enlightenment of modern scientific techniques. In this new age of ultramicroscopic exploration, we shall be getting up close and personal to the viruses themselves. As we shall discover, viruses really are far more embedded in us, and in the biological and ecological world that we inhabit, than most ordinary folk might possibly imagine.

I

What Are Viruses?

Only in the last decade have we come to realise that, from its very beginnings, all of cellular life has inhabited not only the visible biosphere – of solid earth, air and oceans – but also a less familiar and invisible virosphere. The viruses that constitute this virosphere are not merely surrounding us, they are within us, both as evolving extrinsic organisms in themselves and as interactive symbiotic entities that are an intrinsic part of our being. We might not be aware of the presence of these minuscule passengers within us from moment to moment, but the passengers are, in their quintessentially viral way, aware of us.

This might seem somewhat daunting, even frightening, to some of us, but there is no need for alarm. They have always been there. It is likely that they preceded any origins of human life on planet Earth, or indeed, going further back, the origins of the mammals, or any animals or plants, or fungi, or, if I am right, even the single-celled amoebae. All that has changed is that the world of virology is coming to understand the role of viruses in the origins and diversity of life, and what might appear incongruous to any notion of viruses being exclusively agents of disease, the health of the biosphere.

For viruses to achieve all this they must surely possess some remarkable properties. For example, they have no means of loco-motion yet they move among us: in pandemic forms they effortlessly circulate around the globe. Although they have no sense of vision, hearing, touch, smell or taste, they detect with uncanny precision the cell, or organ or tissue that is their target destination. This they achieve despite the relentless opposition of powerful immune defences designed to prevent this happening; and once arrived, they penetrate the defences of the target cell, break entry through its protective surface membrane, and once inside take over its physiological, biochemical and genetic program-ming to compel the cell to become a factory for the production of a new generation of themselves.

Welcome to the world of viruses!

It is, admittedly, a very strange world replete with mysteries. It becomes all the more quixotic when we attempt to examine it at the most basic level.

What then are viruses? How do we even begin to define them? What, for instance, is the difference between, say, a bacterium and a virus? While viruses and bacteria are often confused in the minds of ordinary folk, because they cause many of the common infectious diseases, bacteria and viruses are very different entities. Viruses are more difficult to define than bacteria because they are said to occupy a position somewhere between the biological notions of life and non-living chemicals. This has tempted a distinguished colleague to dismiss them as 'a piece of mischief wrapped up in a protein'. While the hubris contains a grain of truth, there is rather more to viruses than being merely a source of mischief. So let us delve a little deeper! Do viruses rely on genes, and genomes, like all of the more familiar forms of life, from whales to humans and buttercups to the so-called 'humble' bacterium? The answer to that question is, 'Yes!' Viruses do indeed have genomes, which contain protein-coding genes. We shall discover more about those viral genomes in subsequent chapters when we shall also observe some important differences between the genomes of viruses and all other organisms.

Do viruses follow the same patterns of evolution as, say, plants and animals? The answer again is, 'Yes!' But the patterns of evolution – the specific mechanisms involved – are heavily influenced by a facet of their organismal existence that is confined to viruses. Viruses can only replicate by making use of the host cell's genetic apparatus, and because of this, viruses were formerly defined as 'obligate genetic parasites'. But with our increasing understanding of viruses, and of their complex roles in relation to the evolution of their hosts, this definition is no longer sufficient to characterise them. A more adequate definition must take on board the fact that viruses are symbionts. Indeed, we now know that viruses are the ultimate symbionts, exhibiting many examples of all three patterns of symbiotic behaviour, namely parasitism, commensalism and mutualism. Moreover, since viruses will sometimes employ aggression as an evolutionary pattern of behaviour in relation to their hosts, they are also potentially 'aggressive symbionts'.

The more we examine viruses, in their evolutionary trajectories and in the influence of that trajectory on the evolution of their hosts, the stranger and more fascinating their story becomes. Is it reasonable to propose that viruses were born at the stage of chemical self-replicators before the actual advent of cellular life on Earth? If so, how then, from those primal beginnings, did viruses evolve, to interact with, and thus contribute to, the evolution of all of life on this planet?

The aim of this book will be to enlighten readers through a step-wise progression, starting with a familiar territory: we shall confront the wide range of illnesses that are caused by viruses. For example, we shall examine what is really going on in the common cold, the childhood illnesses such as measles, chickenpox, herpes and mumps, rubella, as well as less familiar examples such as rabies, 'breakbone fever', haemorrhagic fevers such as Ebola, and virus-induced cancers such as Burkitt's lymphoma. In such an examination we shall discover what makes viruses tick, exploring what's actually going on inside us when we encounter the virus, how this gives rise to the symptoms we get from the infection, and, key to deeper understanding, probing what the virus itself gets from the 'interaction' with its human host. We shall employ the same virus-orientated perspective to explore important epidemic forms such as influenza, smallpox, AIDS and polio, which will illustrate how viral infections have impacted on human social history, from the wall paintings of the Ancient Egyptians to the colonisations of the Americas, Australasia and Africa. We shall also take a close look at vaccines as a measure to prevent epidemic infections, from the first introduction of vaccination against smallpox centuries ago to the recent controversy concerning the triple vaccine and the papilloma virus vaccine.

The science of virology grew out of the study of viruses in the causation of disease. Through understanding the viruses already familiar to us, we shall widen our enlightenment by examining the role of viruses in the evolution of life, and in particular we shall explore the role of viruses in our human evolutionary history. We shall see how, throughout our prior evolution, we have shared our existence with these powerful invisible entities, and how they really have changed us at the most intimate level, to help make us human.

I hope that, like me, you will come to appreciate the enormous importance of viruses to life, in its origins and complexity, while also marvelling at the existential nature of what is one of the great wonders of life on our beautiful blue-oceanic planet. Viruses, by and large, have had a bad press. This is understandable, given the experiences of earlier generations of virologists, whose only contact with viruses was in dealing with the infections caused by them. But today a major wind of change is blowing through the world of virology – so much so that recently a distinguished evolutionary virologist declared that we were witnessing what he called 'The Great Virus Comeback'. What does he mean by this? Why have some of the modern pioneers of virology introduced the term 'virosphere' as the key to a new exploration of the importance of viruses to the entire biosphere? Could it be true that, as some would have us believe, viruses should now be seen as the 'Fourth Domain of Life'?

2

Coughs and Sneezes Spread Diseases

Historically, viruses were included with the so-called 'microbes' – tiny organisms that were originally discovered as the cause of infectious diseases in humans, animals and plants. Interestingly, there is a part of us that has long been intimately acquainted with microbes in general, and with viruses in particular. This is our inbuilt system of defences against infection: what doctors refer to as our immune system. It is perhaps as well that we possess this inbuilt immunological protection, because we inhabit a world that teems with microbes.

A veritable zoo of such microbes covers our skin and other surface membranes. Biologists call this the 'human microbiome'. Although it might cause some of us to squirm a little just to acknowledge its existence, this secret world is no real threat to us. It is an intrinsic part of our being, comprising a variety of bacteria, as well as other microbial forms, that inhabit our surface skin, mouth and throat, nostrils and nasal cavities, and in the case of women, the genital passages. Our bodies are said to contain roughly 30 to 40 trillion cells – if you are mathematically inclined, this is 3 to 4 times 10^{13} – which comprises the sum total of living cells that make up our living tissues and organs. Meanwhile our 'microbiome', which amounts to all of the microbes that inhabit our skin, gut, oral and nasal passages and throat, and genital tract in women, accounts for some ten times as many microbial cells, comprising such organisms as bacteria, Archaea and protists. It is natural enough, given our awareness of past epidemics and day-to-day troublesome infections, to assume that such microbes are invariably harmful; but these microbes that make up our personal microbiomes are not hostile. Some simply live off us in commensal fashion without causing us any harm; while many others help to maintain our normal health. For example, the zoo of microbes that inhabit our large bowels, or colons, play an important beneficial role in our human nutrition – such as in helping us to absorb vitamin B12 – as well as helping to protect us from invasion of our digestive tract by pathological visitors. The bodies of this 'colonic flora' account for no less than 30 per cent of the bulk of our faecal waste.

There is also growing evidence that we benefit in a number of other ways from this microbial flora of our skin, and other abdominal cavities. This holistic realisation begs a relevant question: could viruses be a part of this human biome, capable of contributing to our human health? For any group of microbes to contribute to the nutrition or general well-being of a host, this would imply a lengthy period of symbiotic evolution with the same host. Immediately we even come to consider such a curious virus–host interaction, we are obliged to consider something profoundly different about viruses when compared to cellular symbionts, such as the bacterial flora of the human intestine or skin. Viruses inhabit the landscape of the host genome.

This means that viruses are certainly not going to contribute, for example, to human vitamin digestion. What it really implies is that, if viruses are to contribute in some way to host health – or indeed host evolution – that contribution is likely to be much more subtle, involving, perhaps in the human host, an interaction with our immunological defences, or more profoundly still, an interaction with our human genetic machinery – or most profound of all, changing our very human genome, the repository of our human heredity, buried deep in the nucleus of every human cell. If this were to happen, viruses would have contributed to what makes us human.

These are weighty questions. Perhaps many of my readers might be inclined to make the point that, so far as they are aware, only the less helpful kinds of viruses appear to have come their way.

In this book we shall explore the truly strange, and intriguing, world of viruses. We might make a start by dispelling a common misconception; many people tend to confuse viruses with bacteria. This is perfectly understandable since viruses, like bacteria, cause many of the common ailments that afflict us in our ordinary lives, and particularly so the fevers that beset the lives of our children. Family doctors deal with these common ailments on a day-to-day basis, and they tend to treat them in similar ways, with antibiotics for bacterial illnesses and vaccination programmes or antiviral drugs aimed at protecting kids from the common viral infections. It is little wonder that people are apt to confuse viruses with bacteria. What then is the difference between the two?

In fact there are major differences between bacteria and viruses. The most obvious difference is one of scale: most viruses are much smaller than bacteria. We readily grasp this if we take a closer look at what is going on during those coughs and sneezes that we recognise as the harbingers of that bothersome cold. While a few other viruses can cause an illness resembling a cold, the majority of colds are caused by a particular virus, which goes by the name of 'rhinovirus'. If one harks back to the sneezing, snuffling and nose-blowing that are the familiar symptoms of that developing cold, the name rhinovirus is apt, since 'rhino' derives from the Greek word, *rhinos*, for nose. Rhinoviruses are the commonest virus infections to afflict humans worldwide, with a seasonal peak in the autumn and early winter. The more we learn about the rhinovirus, the more we witness how well-suited it is to its natural environment, and to its life cycle of infectious behaviour and spread.

The rhinovirus is exceedingly tiny, at about 18 to 30 nanometres in diameter. A nanometre, or nm, is one-thousand-millionth of a metre. This clearly tells us that a single rhinovirus organism – it is referred to as a 'virion' – is absolutely minuscule. In the evolutionary system of classification known as 'taxonomy', rhinoviruses are classed as a genus within the family of the 'picornaviruses', a word derived from *pico* for small, and *rna*, because the rhinovirus genome is made up of the nucleic acid RNA rather than the more familiar DNA. Let us put aside any discussion of these genetic molecules for the moment, but we shall return to consider some remarkable implications of RNA-based viral genomes in subsequent chapters.

Returning to the differences in scale between viruses and bacteria, rhinoviruses are far too small to be seen under the ordinary laboratory light microscope. The virions can only be visualised under the phenomenal magnification of the electron microscope, when they appear to be roughly spherical in shape, resembling tiny balls of wool. In fact, if we examine the individual virions more closely under the electron microscope, we see that they are not really spheres but have multifaceted surfaces, rather like cut diamonds. In the technical jargon, the multifaceted surface of the rhinovirus is the viral 'capsid', which is the viral equivalent of a human cell's enclosing membrane. This capsid has a striking mathematical symmetry comprising 20 equilateral triangles. All viruses have genomes, made up of either DNA or its sister molecule, RNA. The protein capsid acts as a protective shell that encloses the viral genome. It is the capsid that gives rhinoviruses their quasi-crystalline appearance, known as 'icosahedral' symmetry – the term is simply the Greek for 'twenty-sided'. The multifaceted symmetry is not comprised of diamantine crystal, however, but constructed by a biochemical protein assembly.

Microbiologists had long recognised the presence of viruses before the electron microscope was invented. They found ways of detecting the presence of viruses from their effects on host cells, and they could even count their precise numbers from their cytopathic effects in cultures. It will come as no surprise to discover that the best cultures for growing rhinoviruses are cells derived from the human nasal lining, or the lining of the windpipe, or trachea. We are equally unsurprised to learn that the best temperature at which to culture cold viruses is between 33°C and 35°C, which is the temperature found within our human nostrils on a cold autumnal or winter's day.

Rhinoviruses are highly adapted for survival in their host environment. They are also highly adapted to infect a specific host. This became apparent when scientists attempted to infect laboratory animals, including chimpanzees and gibbons, with a variety of different subtypes of rhinovirus that readily infected humans. They could not replicate the symptoms of a typical cold in any of the animals. From this we glean an important lesson about viruses: the rhinovirus is most particular when it comes to its choice of host, which is exclusively *Homo sapiens*. This has a pertinent significance; it means that human infection is vitally

important to virus survival. Only through human to human contagion can the virus be passed on and breed new generations of
rhinovirus. We are the natural reservoir of the cold virus.

But a moment or two of reflection on such exclusivity provokes
a tangential thought – and a pertinent question. These minuscule
polyhedral balls have no obvious means of locomotion. How can
they possibly move about through our human population to effortlessly spread their infection across national and even international
boundaries?

In fact, we already have the answer: it is implied in the very
title of this chapter. Why do we cough and sneeze? We do so
because this is what happens when our noses, throats and windpipe passages feel irritated. It is part of the natural defences against
foreign material entering passages where it could block our airways
and, implicitly, obstruct them and threaten our breathing. What
rhinoviruses do is to provoke the same physiological responses by
irritating the linings of our nasal passages. The viruses spread from
person to person because they are explosively ejected into the
ambient air with every cough and sneeze, to be inhaled and
subsequently infect new hosts. And here, once again, we learn
something vitally important about viruses. The viruses do not
need any mechanism of locomotion because they hitch a ride on
our own locomotion, and everywhere we go, we further oblige
them by spreading their contagion by coughing and sneezing.

How clever, we are inclined to think, are viruses!

But viruses could not possibly be clever. They are far too simple
to be capable of thinking for themselves. We are instead confronted
by another of the numerous enigmas in relation to viruses. How,
for example, could an organism some paltry 30 nanometres in
diameter acquire such devious but also such highly effective
patterns of behaviour as we discover in the common cold? The

answer is that viruses do this through their evolution. Indeed, viruses have an extraordinary capacity to evolve. They evolve much faster than humans, even much faster than bacteria. Over subsequent chapters we shall see how that viral employment of host locomotion is one of many such evolutionary adaptations.

What then do rhinoviruses do when they get inside us?

We have seen that the rhinovirus has a specific target cell, the cilia-flapping cells lining the nasal passages. Once inhaled, the virus targets these lining cells, discovering a specific 'receptor' in the cell's surface membrane, after which the virus uses the receptor to break through the membranous barrier and gain entry into the cell's interior, or cytoplasm. Here the virus hijacks the cell's metabolic pathways to convert it into a factory for the replication of daughter viruses. The daughter viruses are extruded into the nasal and air passages, there to search out new cells to infect and continue the invasive process. It seems to require only a tiny dose of virus to be inhaled from the expelled cough or sneeze of an infected person to initiate infection in a new individual. After arrival, the incubation period from virus entry to infected nasal cells exuding new daughter viruses can be as little as a day. We don't have much of a chance of escaping infection once the virus has been inhaled. Virus replication peaks by day four.

Fortunately, it isn't all one way. Even as the virus is launching its attack, the human immune system has registered the threat, and it has recognised the viral antigenic signature – what we call the scrotype. The problem is that the arrival of a *new* serotype requires time for the immune system to recognise the threat and to build up a formidable arsenal of responses. By day six the nasal passages are the focus of a virus versus immunological war zone, with no quarter asked or given on either side. This intense immune response causes the nasal passages to shed most of their lining cells, exposing highly inflamed raw surfaces, with narrowed breathing passages exuding copious mucus, which contains rising levels of antibodies to the virus. The rhinovirus is eventually killed off by the neutralising antibodies and the 'war detritus' is cleared away by the gobbling action of phagocytic white cells. During this immunological conflagration the new host follows the same unfortunate cycle of being infectious to others, through coughing and sneezing, for a period of anything from one to three weeks.

There is an adage that colds will not kill you. This is largely true, but colds can make children more liable to sinusitis and otitis media, a nasty bacterial infection of the middle ear. Colds can also precipitate asthma in people constitutionally prone to it, and they can provoke secondary bacterial chest infections in people suffering from cystic fibrosis or chronic bronchitis. Nevertheless, the salutary consolation is that, in the great majority of human infections, the rhinovirus eventually passes on by and we make a complete recovery.

Is there anything we can do to minimise the risk of contracting that cold – or is there any effective treatment when we are afflicted?

In Roman times, Pliny the Younger recommended kissing the hairy muzzle of a mouse as a remedy for colds. Benjamin Franklin was more sensible, suggesting that exposure to cold and damp in the atmosphere was responsible for the development of a cold. He also recommended fresh air and avoiding the exhaled air of other people. More modern times have seen a veritable cornucopia of quack remedies for prevention or treatment of colds. One of the most popular was vitamin C, championed by the distinguished American chemist, Linus Pauling. But alas, when subjected to scientific scrutiny it proved no more effective than the mouse's whiskers. Perhaps we should focus more on common sense? Colds are contracted from the coughs and sneezes of infected people.

People in congested offices, or even relatives who find themselves ill at home, should follow the old adage: trap your germs in a handkerchief. If an individual is deemed to be at a particularly high risk from a cold, wearing a virus-level face mask would certainly reduce the likelihood of infection when exposed to an infectious source.

A pertinent question remains: why, then, if our immune system has come to recognise and react to the rhinovirus, are we still susceptible to further colds during our lifetime? In fact, there are roughly 100 different 'serotypes' of the rhinovirus, so immunisation to any one type would not provide adequate protection from the others. Added to this is the fact that serotypes are capable of evolving so that their antigenic properties are apt to change.

3

A Plague Upon a Plague

In 1994 the East African nation of Rwanda erupted onto the
world's news and television screens when a simmering civil war
between the major population of Hutus and minority population
of Tutsis erupted into a genocidal slaughter of the minority population. But despite the deaths of half a million Tutsis, the Hutu
perpetrators lost the war, causing more than two million of them
to flee the country. Roughly half of these fled northwest, across
the border of what was then Zaire, these days the Democratic
Republic of the Congo, where they ended up around the town of
Goma. Up to this point Goma had been a quiet town of some
80,000 people, nestling by Lake Kivu in the lee of a volcano.
Goma now found itself overwhelmed by a desperate torrent of
refugees, carrying everything from blankets to their meagre rations
of yams and beans. Two hundred thousand arrived in a single day,
confused, thirsty, hungry and homeless. They camped on door-
steps, in schoolyards and cemeteries, in fields so crowded that
people slept standing up. Agencies from the world's media flocked
to the vicinity, reporting the chaos and the urgent need for shelter,
food and water.

A reporter for *Time* magazine estimated that the volume of

refugees needed an extra million gallons of purified water each day to prevent deaths from simple thirst, meanwhile the rescue services were managing no more than 50,000. Desperate people foraged for fresh water, scrabbling hopelessly in a hard volcanic soil that needed heavy mechanical diggers to sink a well or a latrine. Human waste from the relief camps fouled the waters of the neighbouring Lake Kivu, creating the perfect circumstances for the age-old plague of cholera to emerge. Within 24 hours of confirmation of the disease some 800 people were dead. Then it became impossible to keep count.

Viruses are not the only cause of plagues, which include a number of lethal bacteria, such as the beta-haemolytic streptococcus, tuberculosis and typhus, as well as some protists, which cause endemic illnesses such as malaria, schistosomiasis and toxoplasmosis. Cholera is a bacterial disease, caused by the comma-shaped *Vibrio cholerae*. The disease is thought to have originated in the Bengal Basin, with historical references to its lethal outbreaks in India from as early as 400 CE. Transmission of the germ is complex, involving two very different stages. In the aquatic reservoir the bug appears to reproduce in plankton, eggs, amoebae and debris, contaminating the surrounding water. From here it is spread to humans who drink the contaminated water, where it provokes intense gastroenteritis, which proves rapidly fatal from massive dehydration as a result of the fulminant 'rice-water' diarrhoea. This human phase offers a second reservoir for infection to the bug. If not prevented by strict hygiene measures, the extremely contagious and virulent gut infection causes massive effluent of rice-water stools that are uncontrollable in the individual sufferer, so that they contaminate their surroundings, and especially any local sources of drinking water, leading to a vicious spiral of very rapid spread and multiplication of the germ.

During the nineteenth century, cholera spread from its natural heartland, provoking epidemics in many countries of Asia, Europe, Africa and America. The massive diarrhoeal effluent of cholera is unlike any normal food poisoning. An affected adult can lose 30 litres of fluid and electrolytes in a single day. Within the space of hours, the victims go into a lethargic shock and die from heart failure.

The English anaesthetist, John Snow, was the first to link cholera with contaminated water, expounding his theory in an essay published in 1849. He put this theory to the test during a London-based outbreak around Broad Street, in 1854, when he predicted that the disease was disseminated by the emptying of sewers into the drinking water of the community. Snow's thoughtful research led to the civic authorities throughout the world realising the importance of clean drinking water. Today the life of an infected person can be saved by very rapid intravenous replacement of fluid and electrolytes, but the size of the outbreak around Lake Kivu, and the relative paucity of local medical amenities, limited the clinical response. The situation was made even worse by the recognition that the cholera in the Rwandan refugee camps was now confirmed as the 01-El Tor pandemic strain of *Vibrio*, known to be resistant to many of the standard antibiotics. This presented immense problems for the medical staff from local health ministries and those arriving from the World Health Organization. Even though the response was one of the largest relief efforts in history – involving the Zairian armed forces, every major global relief agency and French and American army units – the spread of cholera was too rapid for their combined forces to take effect.

Three weeks after the outbreak began, cholera had infected a million people. Even with the modern knowledge and the desperate efforts of civic and medical assistance, the disease is believed to have killed some 50,000. It is hard to believe that so resistant a plague bacterium as the *Vibrio cholerae* might itself be prey to another microbe. But exactly such an attack, of a mystery microbe upon the cholera vibrio, had been recorded in a historic observation by another English doctor close to the very endemic heartland of the disease, a century before the outbreak at Lake Kivu.

In 1896 Ernest Hanbury Hankin was studying cholera in India when he observed something unusual in the contaminated waters of the Ganges and Yamuna Rivers. Hankin had already discovered that he could protect the local population from the lethal ravages of the disease by the simple expedient of boiling their drinking water before consumption. When, in a new experiment, he added unboiled water from the rivers to cultures of the cholera germs and observed what happened, he was astonished to discover that some agent in the unboiled waters proved lethal to the germs. It was the first inkling that some unknown entity in the river waters appeared to be preying upon the cholera bacteria.

Hankin probed the riddle further. He found that if he boiled the water before adding it to the cholera germ cultures this removed the bug-killing effect. This suggested that the agent that was killing the cholera germs was likely to be of a biological nature. He needed to know if it was another germ – sometimes germs antagonised one another – or if it was something completely different, a truly mysterious agent, that was killing the germs. Hankin decided that he would set up a new experiment using a device known as a Chamberland-Pasteur 'germ-proof' filter, which had been developed 12 years earlier by the French microbiologists Charles Chamberland and Louis Pasteur. The Chamberland-Pasteur filter was a flask-like apparatus made out of porcelain that allowed microbiologists to pass fluid extracts through a grid of pores varying from 0.1 to 1.0 microns in diameter – from 100-billionths to 1,000-billionths of a metre – that were designed to trap bacteria but allow anything smaller to pass through. Two years after the filter's invention, a German microbiologist, Adolf Mayer, showed that a common disease of tobacco plants, known as tobacco mosaic disease, could be transmitted by a filtrate that had passed through the finest Chamberland-Pasteur filter. Unfortunately, he persuaded himself that the cause of the disease must somehow be a very tiny bacterium. In 1892 a Russian microbiologist, Dmitri Ivanovsky, repeated the experiment to get the same results. He refuted a bacterial cause, but still arrived at the mistaken conclusion that there must be a non-biological chemical toxin in the liquid extract. Finally, in 1896, the same year that Hankin was looking for his mystery agent in the Indian river waters, a Dutch microbiologist, Martinus Beijerinck, repeated the filter experiment with tobacco mosaic disease; but Beijerinck concluded that the causative agent was neither a bacterium nor a chemical toxin but rather 'a contagious living fluid'. Although Beijerinck was closest of all to the truth, he was once again wrong. Today we know that the cause of tobacco mosaic disease is a virus – the tobacco mosaic virus. But thanks to Beijerinck's mistaken finding of a 'contagious fluid', the current Oxford English Dictionary definition of a 'virus' has it as: 'a poison, a slimy fluid, an offensive odour, or taste'.

Viruses are not poisons, or slimy fluids, or offensive odours or tastes, but rather organisms – truly remarkable organisms – that are different from bacteria, indeed utterly different from any other organisms on Earth. The great majority of viruses are very small, tiny enough to pass through Chamberland-Pasteur filters.

Of course, Hankin knew nothing of the existence of viruses when he passed the river water through the refined sieve of a Chamberland-Pasteur filter. Although he was in no position to offer a likely explanation or name for the mystery agent, he had discovered one of the most important and ubiquitous of viruses on Earth: a member of the group known today as 'bacteriophage' viruses, so-named from the Greek *phagein*, which means to devour. That is exactly what was happening to the cholera germs in Hankin's experiments: they were being 'devoured' by bacteriophage viruses.

The true nature of Hankin's discovery remained a mystery until 1915, when English bacteriologist Frederick Twort discovered a similarly minuscule agent that could pass through the Chamberland-Pasteur filters and yet remained capable of killing bacteria. By now viruses were known to exist even though biologists knew very little about them. Twort surmised that he was observing either a natural phase of the life cycle of the bacteria, the result of a fatal enzyme produced by the bacteria themselves, or a virus that grew on and destroyed the bacteria. Some two years later, a pioneering, self-taught, French-Canadian microbiologist, Félix d'Herelle, finally solved the mystery.

D'Herelle was born in the Canadian city of Montreal but considered himself a citizen of the world. Before becoming involved with viruses, he had already travelled widely, working in numerous American, Asian and African countries, to finally settle at the Pasteur Institute in Paris. At this time the discipline of microbiology was a fashionable scientific research endeavour and it was rapidly expanding its knowledge base. During his researches in Tunisia, d'Herelle had come across what was probably a virus infecting a bacterium that itself caused a lethal plague in locusts. Now working at the famous Institute, even as the First World War raged nearby, he took a particular interest in the grimy disease known as bacterial dysentery, which was killing soldiers in their muddy trenches.

Bacterial – as opposed to amoebic – dysentery is caused by a genus called *Shigella*, which is passed on from the infected individuals through faecal hand-to-mouth contagion. The resultant illness ranges from a mild gut upset to a severe form, with agonising griping spasms of the bowel accompanied by high fever, bloody diarrhoea and what doctors call 'prostration'. In July and August 1915 there was an outbreak of haemorrhagic bacterial dysentery among a cavalry squadron of the French army, which was stalemated on the Franco-German front little more than 50 miles from Paris. The urgent microbiological investigation of the outbreak was assigned to d'Herelle. In the course of intensive investigation of the bugs responsible, he discovered 'an invisible, antagonistic microbe of the dysentery bacillus' that caused clear holes of dissolution in the otherwise opaquely uniform growth of dysentery bacteria on agar culture plates. Unlike his earlier colleagues, he had no hesitation in recognising the nature of what he had found. 'In a flash I understood: what caused my clear spots was . . . a virus parasitic on the bacteria.'

D'Herelle's hunch proved to be correct. Indeed, it would be d'Herelle who would give the virus the name we know it by today: he called it a 'bacteriophage'. Then the French-Canadian microbiologist had an additional stroke of luck. When studying an unfortunate cavalryman suffering from severe dysentery, he performed repeated cultivations of a few drops of the patient's bloody stools. As usual, he grew the dysentery bug on culture plates and passed a fluid extract through a Chamberland-Pasteur filter, thus obtaining a filtrate that could be tested for the presence of virus. Day after day, he tested the filtrate by adding it to fresh broth cultures of the dysentery bug in glass bottle containers. For three days the broth quickly turned turbid, confirming teeming growth of the dysentery bug. On the fourth day new broth cultures initially became turbid as usual, but when he incubated the same cultures for a second night he witnessed a dramatic change. In his words, 'All the bacteria had vanished: they had dissolved away like sugar in water.'

D'Herelle deduced that what he was witnessing was the effects of a bacteriophage virus, which must also be present in the cavalryman's gut – a bacteriophage virus that was capable of devouring the *Shigella* germ. But then he had an additional stroke of genius. What if the same thing was happening inside the infected patient? He dashed to the hospital to discover that during the night the cavalryman's condition had greatly improved and he went on to make a full recovery. At this time bacterial infections, such as dysentery, typhoid fever, tuberculosis and the streptococcus, were a major cause of disease and death throughout the world. With no known antibiotics to treat infections, there was a desperate need for any form of therapy. His observations with the dysentery bug bacteriophage gave d'Herelle the idea that, perhaps, phage viruses might be cultivated with the express purpose of treating dangerous bacterial infections.

During the 1920s and 1930s, d'Herelle conducted extensive research into the medical applications of bacteriophages, introducing the concept of phage therapy for bacterial infections. This therapy saw widespread use in the former Soviet Republic of Georgia, and also the United States, continuing in use until the discovery of antibacterial drugs in the 1930s and 1940s. The use of drugs was much simpler to apply and proved dramatically effective, thus supplanting bacteriophage therapy. But this did not stop d'Herelle from continuing to study this marvellous if deadly entity that was so very tiny that it was completely invisible even to the most powerful light microscope, and yet appeared to be so powerful when it came into contact with its prey bacteria.

In 1926, d'Herelle published a now-historic book, *The Bacteriophage*, in which he described his work, and thoughtful extrapolations, concerning bacteriophage viruses. As we shall duly discover, the importance of the bacteriophage, as we recognise it today, has eclipsed all that even its pioneering researcher, Félix d'Herelle, could possibly have imagined in those early years.

In retrospect, it is remarkable that, even so many decades ago, d'Herelle clearly grasped that he was dealing with a wonder of the natural world, declaring in his book that these agents that were so dreadfully lethal to bacteria were also capable of exerting an extraordinary balancing effect in the interactions between the bacteriophage virus and its host bacterium. In his words: 'A mixed culture results from the establishment of a state of equilibrium between the virulence of the bacteriophage corpuscles and the resistance of the bacterium. In such cultures a *symbiosis* obtains, in the true sense of the word: parasitism is balanced by the resistance to infection.' This is the first use of the term 'symbiosis' in reference to viruses in microbiological history. In a footnote, d'Herelle took the implications further by drawing a parallel between what he was observing in the interaction of the bacteriophage virus and bacterium and the symbiosis that had recently been discovered in all land plants, where fungi in soil invade the plant roots to form a 'mycorrhiza', whereby the fungus feeds the plant with water and minerals and the plant feeds the fungus with the energy-giving metabolites that derive from the photosynthetic capture of sunlight. In d'Herelle's words: 'The respective behaviour between the bacterium and the bacteriophage is exactly that of the seed of the orchid and the fungus.'

D'Herelle is now recognised by many scientists as the father of both virology and molecular biology. But it would take many years before the world of virology, and microbiology in general, would come to rediscover d'Herelle's original vision of the symbiotic nature of the bacteriophage.

4

Every Parent's Nightmare

Parents will be familiar with the anxiety that comes with childhood rashes and fevers. How natural that our hearts should falter with the beloved child tossing in a perspiring fever, the restless anxiety, racking coughs, or sickness and vomiting. We can hardly sleep with worry that something worse might happen in the dark of night. That worry is, perhaps, a residuum of a fear from times only recently gone by when unpleasant things really did happen in the dark of night to those we loved. How fortunate we are now that our families are protected by antibiotics, antiviral drugs and the vaccines that keep such terrors at bay. But these advances are relatively new to medicine and to society. We should not forget that as recently as the 1950s most of humanity, even in developed countries, ultimately died from infection.

Before its prevention, using the triple vaccine, one such major cause of parental anxiety was measles, a commonplace and highly contagious childhood fever. How astonishing it is that this appears to be a relatively new disease in humans. Hippocrates, who wrote about the common diseases in Ancient Greece in the fifth century

CE, recorded clearly recognisable descriptions of common infections such as the virus-caused herpes and the protist-caused malaria, yet this very knowledgeable ancient authority gave no description to match the symptoms and signs of measles. It is hardly a disease that would be readily missed, with its striking rash and fever, high contagion and common association with childhood. There is a clue in the name, 'measles', deriving from an Anglo-Saxon word *maseles*, which means 'spots'. The first written description of measles is attributed to the tenth-century Persian physician, Abu Becr, also known as Rhazes, who cited a seventh-century Hebrew physician, El Yehudi, as providing the first clinical description of the disease. Rhazes recognised measles as an affliction of children and he distinguished it from the equally prevalent but far deadlier rash-provoking disease of smallpox.

Symptoms typically include a high fever, with a temperature often greater than 40°C, a racking cough, runny nose and inflamed eyes. Two or three days after the start of the fever, small white spots on a red inflamed background can be seen in the mucous membranes inside the cheeks. These are known as Koplik's spots and are diagnostic of the disease. At much the same time a flattish, bright red rash invades the skin, usually beginning on the face and then spreading to the rest of the body. The rash, and causative illness, usually persists for seven to ten days and, in fit and well-nourished children, is usually followed by a full recovery. But in a minority of cases, most commonly seen in malnourished children, and in particular children in less-developed countries with poorly developed health care facilities, measles can lead to serious complications.

Like the common cold, measles is specific to humans, although it can be artificially transmitted to monkeys by laboratory experiment. This means that we are the reservoir of measles in nature

– we are the natural host. The only place measles virus can spread its infection, and produce its new brood of new daughter viruses, is in us. It really is that up close and personal. And this means that the relationship – the symbiosis – between humans and the measles virus has been evolving for a long time, and in symbiological parlance with evolutionary implications for both 'partners'. The causative virus, or morbillivirus, comes in a variety of groups, known as 'clades', within the broader family of viruses, called the paramyxoviruses. Individual measles virions are spherical, rather like cold viruses, with a genome made up of a single strand of RNA. The viral genome is contained within a similar capsid type of coat, but in this case the capsid is enclosed in an additional surface 'envelope', which carries multiple spike-like projections that play a key role in the infectious process.

Measles is a highly infectious virus with a worldwide distribution, but it can only survive in populations as an 'endemic' contagion, in populations where there is a continuous supply of susceptible children. We shall return to this observation when talking about the measles vaccine. The measles virus spreads by aerosol inhalation, much like the common cold. Its initial target cells are, once again, the lining cells of the respiratory tract. But unlike the cold virus, with its focus on the nose and throat, measles heads down into the lower respiratory tract. For some unknown reason, the virus also has a predilection for the cells of the conjunctivae, which explains the inflamed eyes that are a common sign of the clinical presentation. During the first two to four days of infection, the virus multiplies locally in the target cells. The alien presence of the virus provokes local inflammation and this in turn attracts the attention of white blood cells, known as macrophages, which normally gobble up unwanted debris, dead and diseased cells and invading parasites. This process is known as phagocytosis.

Unfortunately – or alas perhaps knowing a little about viruses and their behaviour, predictably – these same phagocytes now become the final target cells of the measles virus.

The virus hijacks the phagocytes, invading and then replicating inside them, and then taking advantage of their natural locomotion to the regional lymph glands, where a second phase of viral replication takes place. From the lymph glands, the virus invades another variety of white blood cells, known as leukocytes, and once again it hitches a ride aboard these infected cells into the bloodstream, thus spreading to every cell and tissue, notably the skin. It is at this stage of bloodstream spread, or 'viraemia', that the typical rash and high fever appear.

Just as we saw with the cold virus, the measles virus doesn't have things all its own way. Those same cells targeted by the multiplying virus, the macrophages, are the first line of defence in our immune system. Besides phagocytosis, the macrophages play a critical role in our inbuilt 'innate' immunity. They also play a key role in triggering an even more powerful defence system, our 'adaptive' immunity, identifying foreign antigens on the surface membranes of the virus as 'alien' to the body's notion of 'self', and presenting these foreign antigens to cells, such as lymphocytes, that set off a process of specific immune recognition followed by the production of antibodies to the virus. The antibody response is also combined with yet another key element of our immune defences, known as 'cellular immunity'. All of these powerful elements of our immune response will ultimately work together to destroy the foreign threat.

Many years ago, as a medical student at the University of Sheffield, I conducted an experiment aimed at testing how the mammalian immune system would respond to exactly such a viral invasion into our bloodstream. With the help of my mentor, Mike

McEntegart, Professor of Microbiology, I injected viruses into the bloodstream of rabbits and then observed how the rabbit immune system dealt with them. I started with a primary dose and followed this up a week or so later with a booster dose. Some readers might react with concern about hurting experimental animals, but the virus I used was a bacteriophage, known as ΦX174 – a virus that only attacks *E. coli* bacteria – so the rabbits suffered no illness. But their adaptive immune system responded in exactly the way a mammalian immune system should respond to any alien invader entering the bloodstream, with a build-up of antibodies in two waves, rising to a peak by 21 days, by which time a single drop of the now-immune rabbit serum was seen to inactivate a billion viruses in mere minutes. With the help of other colleagues at the university, we obtained pictures of what was actually happening under the electron microscope, which showed the syringe-shaped phage virus being overwhelmed with antibody molecules and gathered up in sticky antibody-wrapped aggregates that would have been readily mopped up and cleared from the system by the ever-vigilant phagocytes.

What I observed in the phage virus experiment is similar to what would be expected to happen in a child suffering from measles. There is an incubation period of one to 12 days after exposure to the virus, during which it is passing through the target cells in the respiratory tract, through the lymph glands and entering the bloodstream. At this stage the illness becomes obvious, with fever, cough, runny nose and inflamed eyes. Two or three days later, the Koplik's spots appear on the inner lining of the cheeks and the rash appears on the face and spreads over a day or two to be confluent over the skin. Ironically the striking symptoms and signs, including the fever and the rash, are actually produced by the attack of the immune system on the virus. Through the actions of that same immune system, the majority of children go on to make a full recovery – after which the immune system retains its memory of the antigens on the surface of the virus. In most cases, this will ensure that the sufferer is resistant to any future infection with measles. But further complications bedevil the recovery in a tragic minority of infected children, which include diarrhoea, pneumonia, blindness and, most serious of all, the inflammation of the brain called encephalitis.

Readers may be astonished to read that before the introduction of the measles vaccine, in 1963, major epidemics of measles swept through the global population every two to three years, causing some 2.6 million deaths. Even today, measles is still one of the leading causes of death in young children, despite the fact that a safe and cost-effective vaccine is available to prevent the infection. Between the years 2000 to 2016, the World Health Organization estimated that measles vaccination had prevented some 20.4 million deaths; but, tragically, in 2016 some 90,000 people still died needlessly from this preventable infection.

Unlike my generation, in which measles infection was commonplace, most parents in developed countries these days will have little or no experience of dealing with measles in the family. This, thankfully, is through the benefit of the MMR vaccination programmes which are now governmental policy in many countries. MMR vaccines protect children against three different viral illnesses: measles, mumps and rubella. But as a result of so-called 'MMR misinformation scares', the triple vaccine has been the subject of controversy in different countries, with some misguided parents withdrawing their children from the vaccination programmes.

I shall return to this important topic later in this chapter, but first I would like to examine the other two viruses involved in the vaccine.

The infection we call 'the mumps' probably derives its name from an old word meaning 'to mope' – an apt description of the afflicted child, struck down by malaise and fever and, a day after the onset, the painful swelling of one or both parotid glands within the cheeks, a condition known clinically as 'parotitis'. The causative virus, the mumps virus, is another paramyxovirus, which is also global in distribution. Unlike measles, mumps was familiar to Hippocrates, some two and a half millennia ago. Mumps is also specific to and dependent on the human host, which, in symbiological parlance, is its co-evolving partner, and sole natural reservoir. Once more, the mumps virus is usually spread by the respiratory route, but it can also be spread through contamination with virus-infected saliva.

Fortunately, in most cases the illness is quickly dealt with by the immune system, with the symptoms settling within a few days, so that recovery is usually uneventful. In some cases the illness is so slight that the sufferer doesn't even realise he or she has encountered the virus. But in 20 per cent of males who contract mumps after the age of puberty, the virus causes inflammation of the testes, clinically known as 'orchitis'. This manifests as local pain, which can be severe, accompanied by the swelling of one or both testes some four or five days after the onset of the parotitis. Though some testicular atrophy may result, thankfully the orchitis doesn't usually cause subsequent sterility. Though uncommon, mumps can occasionally cause inflammation in the ovaries in females, and equally rarely cause pancreatitis in either sex. Mumps may also cause a viral, or 'aseptic', meningitis and, like measles, it may also cause encephalitis. Meningitis and encephalitis are serious medical complications, which will usually result in hospitalisation and, in some cases, mortality.

Rubella, or the so-called 'German measles', is not a German

contagion at all but rather a globally distributed infection. The illness just happened to be first described by two German doctors back in the eighteenth century. No more does it have anything to do with measles. The causative virus is in fact a 'togavirus', and an interesting example of this family of viruses since it is the only togavirus that isn't spread by biting insects. Rubella is a contagious, generally mild, viral infection that mostly afflicts children and young adults. But if the virus infects women in early pregnancy, at a key time when major embryological development is taking place in the foetus, it can cause foetal death or a range of severe congenital defects known as 'congenital rubella syndrome' (CRS). These include hearing impairment, eye and heart defects, autism, diabetes mellitus and thyroid malfunction.

The key fact here is that rubella, like measles and mumps, is exclusive to humans. It means that we are the only reservoir or host of all three viruses – in the symbiological lexicon, we are the exclusive partner. That means that if the reservoir were to be closed down, for example through vaccination, the diseases would disappear.

The risk of all three of these viruses – measles, mumps and rubella – has been greatly reduced in developed countries by preventive vaccination, which, in the UK, the US and many other countries, is achieved using the combined MMR vaccine. It is important, given various misinformation scares, that we grasp the purpose of such a vaccine, and indeed the way in which vaccination works.

Vaccines use either a live, but harmless, variant of a live virus, or a killed virus – or even antigens extracted from parts of a virus – to protect children from the suffering and potential complications of virus infection. The MMR triple vaccine, which employs all three live attenuated viruses – measles, mumps and rubella – has greatly reduced the prevalence of all three viral diseases in the countries where it has been introduced. Unfortunately, a scientifically disproven claim that the MMR vaccine increases the risk of autism has persuaded some parents to forgo vaccinating their children.

People really do need to sit up and take notice of the advice of doctors and health authorities and ignore the misinformation coming from unreliable sources. Not doing so has the potential for unpleasant consequences. In a recent case involving the Somali-American community in the state of Minnesota, the local population, being misguided into believing that the vaccine had increased the frequency of autism in their children, stopped vaccinating their children with MMR. The real truth was exposed by a joint study by the University of Minnesota, the Centers for Disease Control in Atlanta, and the US National Institutes of Health, which showed that the incidence of autism in the Somali-Americans was no different from the vaccinated city's white population. Alas, in May 2017 Minnesota saw the biggest outbreak of measles in the state for 27 years. State officials recommended that the Somali children be protected as soon as possible with vaccination booster shots.

America is far from alone in the resurgence of this dangerous and highly infectious disease of childhood. In May 2018, the British newspaper, the *Daily Telegraph*, reported a resurgence of measles throughout the continent of Europe, with the disease increasing in Belgium, Portugal, France and Germany. Once again, the efficacy of MMR vaccination was being undermined by the same baseless linking of the measles vaccine to autism, which had resulted in a rise from a record low incidence of measles Europewide, with 300 per cent rise in cases from 2017 to an estimated 21,000 cases in 2018, and some 35 reported deaths. In the UK, following years of similar misinformation about a link between the MMR and autism, many people of late teenage years to early twenties had not been vaccinated in their childhood years, making them now susceptible to this unpleasant and potentially dangerous viral infection. In July 2018 *The Times* reported a national alert being sounded out to family doctors throughout the UK, warning them to be on the alert for the disease in families returning from holidays in Italy. In England alone some 729 cases had already been reported in the first half of the year, when compared to 274 in the whole of the previous year.

Parents with any due concerns should seek the advice of their knowledgeable family doctors.

5

A Bug Versus a Virus

One of the commonest errors people make in relation to microbes is to confuse viruses with bacteria. It is important that we recognise the differences since this is the first step towards understanding the vital role of the interactions between the two very different organisms – bacteria and viruses – in the great ecological cycles that are central to life on the planet. One of the commonest of bacterial species found in the healthy colon of mammals is *Escherichia coli*, usually diminished to *E. coli*. The most widely studied bacterium in laboratory experiments, *E. coli* is also an important member of the symbiotic gut bacteria, helping in the production of vitamin K and the digestive uptake of vitamin B12, meanwhile also helping to reduce the threat of invading pathogenic bacteria. *E. coli* colonises the baby's gut within 40 hours of birth, gaining access through hand-to-mouth human contact – most likely the mother during her fondling and feeding of the child. This, of course, is no threat, but rather the beginning of an important symbiotic interaction between human and bacterium.

The *E. coli* species is divided into a number of serotypes, most of which are either harmless or symbiotic to humans. This is why contamination of the skin with human waste is a question of hygiene rather than a cause for alarm. However, there are pathogenic serotypes of *E. coli* that can cause gastroenteritis, and these serotypes may be involved in food scares and product withdrawal from food outlets. More virulent strains of the pathological serotypes can cause urinary tract infections and, rarer still, life-threatening bowel necrosis, peritonitis, septicaemia and fatal cases of haemolytic-uraemic syndrome. Thankfully these serotypes are very rare, so that, under normal circumstances, *E. coli* is a beneficial contributor to the human gut flora.

Under the light microscope, the bacterium is visible as a single-celled sausage-shaped bacterium roughly 2.0 micrometres long. A micrometre, or μm, is one-millionth of a metre. *E. coli* has no nucleus and so it is an example of a prokaryote, which translates from the Greek to mean 'before nucleated life forms'. The bacterial body is enclosed in a membrane, or cell wall, which contains the protein antigens that separate it into different serotypes. The cell wall does not take up the commonly used dye for testing bacterial types, known as a Gram stain, so it is classified as Gram-negative. This same cell wall is capable of acting as a barrier to certain antibiotics, so for example *E. coli* is resistant to the action of penicillin. Many strains of the bug have flagella and so they can be seen to wriggle about in search of nourishment. The bug is attuned to living in the anaerobic environment of the human intestine, where it sticks on tight to the microvilli of the intestinal wall. When passed out of the body, in faeces, the bug is capable of surviving for some time even when exposed to the oxygenated environment. This is why pathological serotypes can cause food contamination in the home and in food-processing environments.

We are somewhat inclined to see all microbes as potential pathogens. But outside the medical world, microbiologists have long been aware that microbes play much wider roles in nature.

For example, the bacteria in soil are essential to the normal cycles of life, helping to break down organic matter to its elemental components, which are then made available for recycling to supply the basic requirements of other living beings. So essential are these soil bacteria that if they were to disappear, the vast majority of life on Earth would follow their example. Such living interdependency is known as symbiosis. We humans are apt to confuse symbiosis with notions of 'friendliness' or 'togetherness', thereby grafting human attributes onto situations where such human notions do not apply. Perhaps it might be a good idea to clarify what the concept of symbiosis actually means to the biological sciences.

Bugs, such as bacteria and viruses, do not think. No more do they have feelings. Their behaviour among themselves, and in relation to their hosts, is driven by a mixture of happenstance and the fundamental mechanisms of evolution. Symbiosis is not about Mr Friendly Guy who shakes the hand of Ms Friendly Lady and everything is hunky-dory from then on. It is about survival in what Darwin called 'the struggle for existence'. In 1878 a professor of botany in Berlin, called Anton de Bary, defined symbiosis as 'the living together of differently named organisms'. A modern interpretation might rephrase his definition as 'living interactions between different species of organisms'. The interacting partner species are called 'symbionts' and the interaction as a whole is called the 'holobiont'.

While symbiosis includes parasitism, which is defined as a symbiotic interaction in which one or more of the partners benefits from the partnership at the expense of another, symbiosis also includes commensalism, where one or more partners gains without detriment to the others; and it also includes mutualism, where two or more of the interacting partners gain from the partnership without harm to the other partner, or partners. It is important to grasp that mutualism often begins as parasitism – indeed in nature many relationships involve situations somewhere between the extremes of parasitism and mutualism. This broader umbrella of living interactions offers the necessary scope for understanding the enormous variety of living interactions, involving microbes and their hosts, in nature. It allows us to compare and contrast a bacterium, *E. coli*, with a virus that also has a predilection for the human gut: the so-called winter vomiting bug, known as the norovirus.

The norovirus is the commonest cause of gastroenteritis in the world, familiar to most of us with its unpleasant manifestations of diarrhoea, vomiting and stomach cramps. It is extremely contagious by the faecal-oral route, whether through contaminated food or water, or direct contact contamination from another sufferer. Once again, we humans appear to be the only host. This, in turn, means that we are the natural reservoir in nature of the virus. Symptoms usually develop some 12 to 48 hours after exposure to infection, often with a low fever and headache. The gut irritation is rarely severe enough to provoke the bloody diarrhoea that is sometimes seen in dysentery, and recovery usually follows within a few days. Since the condition is usually self-limiting, diagnosis tends to be made on the basis of symptoms alone, especially when it occurs during a local recognised outbreak. No specific treatment is usually necessary, although sufferers may be helped by increasing fluid intake to avoid dehydration, together with non-specific anti-fever and anti-diarrhoeal medication. Laboratory confirmation is not usually necessary although public health authorities may sometimes make use of it for contact tracing purposes.

Prevention is the judicious policy, through careful hand-washing and disinfection of potentially contaminated surfaces. Unfortunately, alcohol-based hand sanitisers of the sort dispensed in hospitals are, reportedly, ineffective.

Noroviruses comprise a genus within the family of calciviruses, so-called because they have cup-like depressions in their capsids and so were named after the Greek word *calyx*, which means a cup or goblet. Since they cannot currently be cultured in the usual laboratory media, the single species is divided into six genetically distinct 'genogroups', which infect mice, cows, pigs and humans. The human genotypes are extremely infectious even from minute numbers of the virus, so much so that it has been calculated that a single tablespoonful of diarrhoeal effluent from an infected individual would contain enough viruses to infect everyone in the world many times over. But this is no cause for alarm. Thankfully there is rather more to infectious spread than such theoretical extrapolations. A more practical consideration is the fact that affected individuals can remain infectious for several days *after* the symptoms have settled. This means that they might feel well enough to return to normal life, including work premises, when they are still capable of passing on the virus. It might also contribute to the tendency for outbreaks to occur in closed communities, such as hospitals, cruise ships, schools and residential care homes, where communal food preparation, and common dining areas, make transmission of the virus more likely. Readers may be surprised to learn that, in spite of the relatively mild nature of the illness, the ease of transmission, combined with the prostration of the vomiting and diarrhoea, has led to the norovirus being classed as a Category B bio-warfare agent.

Globally it is estimated that norovirus infects some 685 million people a year, most of whom go on to make a full and speedy recovery. Unfortunately, in a small minority, it can result in a life-threatening illness, with some 200,000 or so deaths worldwide

each year. Children under the age of five years are most susceptible, especially in developing countries, where it causes as many
as 50,000 paediatric deaths annually. It is worrying that the
number of reported outbreaks has been rising since 2002, warning
health authorities, if they weren't sufficiently alarmed already,
that we need to treat the norovirus as a dangerous 'emerging
infection', and one that may be evolving even more highly infectious strains.

The causative virus is globular in shape and between 20 and
40 nanometres in diameter. This means that the norovirus is
somewhere between a hundredth and a fiftieth the size of the
E. coli bacterium. Viruses lack the enclosing cell wall seen in
bacterial, or indeed human, cells. But under the powerful magnification of the electron microscope we see that the norovirus
possesses an icosahedral capsid, which encloses and protects the
viral RNA-based genome. *E. coli*, like all bacteria, and indeed all
cellular forms of life, has a DNA-based genome.

If we compare and contrast the bacterial and viral genomes,
we come across gargantuan differences between bacteria and
viruses at every level of their structure and organisation. The
E. coli genome is coiled into a single, very lengthy circle of DNA
that is attached to the inner aspect of the bacterial cell wall at a
single point. This bacterial genome contains roughly 4,288 protein-
coding genes, as well as coding sequences for other key metabolic
functions involved with the handling of gene expression. This is
comprehensive enough for the bacterium to store the memory of
its genetic heredity as well as to allow it to carry out numerous
internal metabolic functions involved in its internal physiology
and biochemistry. One such key function is the control of the
processes involved in its budding pattern of reproduction, to
produce daughter bacteria.

When compared to the bacterial genome, the norovirus counterpart is frugal in the extreme. The viral genome comprises regulatory regions at either end of a compact linear string of RNA, which codes for a minimum of eight proteins, two of which code for the protein structures of the viral capsid, and six concerned with viral replication. A key difference between the bacterium and the virus is that the bacterium has all it needs to reproduce itself, but the virus can only replicate to produce daughter viruses by making use of the genetic and biochemical properties of its cellular host. In the case of the human strain of norovirus, these genetic and biochemical properties are those of the human target cell.

The norovirus genome codes for a singular aggressive viral protein known as the 'protein virulence factor', or VF1. This menacing entity localises to the human mitochondria during infection with the virus, where it antagonises the infected person's innate immune response to the virus. While some viruses are capable of commensalism or even mutualistic interactions with their hosts, we see little evidence for this in the norovirus. Its symbiotic interaction with humans appears to be exclusively parasitic. Unlike the bacterium, it has no genes devoted to nutrition, or to internal metabolic pathways, since, unlike the bacterium, it has no internal metabolic pathways. Its genome is designed to take advantage of the physiology, metabolic pathways, genetic pathways, and even the very locomotion and life-style patterns of human behaviour in order to replicate itself and transmit its contagion as widely as possible.

So now we see that viruses are not fluids or poisons. They are organisms that follow a wide range of symbiotic interactions, each virus usually associated with a highly specific host, a tiny minority of which happen to be human. They are clearly very different in size, genomic organisation and life-cycle patterns to bacteria. The fact that most viruses do not possess their own internal metabolic processes does not imply that viruses do not utilise metabolic processes. On the contrary, viruses take advantage of their host's metabolic pathways. This is why it is a mistake to think of viruses in isolation from their hosts. Outside their hosts viruses are biologically inactive: but this does not mean that they are inorganic chemicals.

Outside the target cells of their hosts, viruses have evolved stages that are somewhat equivalent to suspended animation. This stage is well-suited to being ejected in the aerosol created by a cough or a sneeze, or excreted in faeces, or in sexual secretions, or surviving being transferred by a secondary carrier, such as a biting insect or a rabid dog; or in the case of plant viruses, being carried to new hosts on the wind, or through water, or through a miscellany of other avenues of transmission, to find new hosts. Only when they enter into their obligate symbiotic partnership with the new host do we witness viruses behaving with the genetic and biochemical subtlety and efficiency we might expect of biological organisms.

The norovirus is no exception to such symbiotic evolutionary behaviour. So specific is the virus in its symbiotic interaction with its human host that different human-associated viral genotypes have affinities for specific ABO blood group proteins on cell membranes, these protein 'receptors' binding with one of the two proteins of the viral capsid as an integral step in the infectious process. On passing into the bowel, the virus has a predilection for the upper small bowel, or jejunum. How, exactly, the virus then penetrates the intestinal wall is not fully understood, but it would appear that it preferentially infects the immune lymphoid follicles in the gut wall, which are known as Peyer's patches, while also searching out a type of intestinal cell, known as H-cells. After making its way through the gut wall, the virus is identified as alien by the innate immune defences of the gut, which might be just fine as far as the virus is concerned, since these may be its target cells. Whatever the target cells, we can anticipate that the virus will hijack their genetic and metabolic pathways in order to replicate itself, thus establishing its cycle of infection and multiplication, generation after generation.

Since we don't yet have suitable tissue cultures or animal models to study the norovirus, we are not in a position to examine the ways in which it provokes the vomiting and diarrhoea, which play a key role in spreading the virus far and wide throughout the world. Currently there is no preventative vaccine, but trials of an oral vaccine are taking place as I write. Let us cross our fingers and hope that these trials are rewarded with an early success!

6

A Coincidental Paralysis

In the summer of 1921 the 39-year-old Franklin D. Roosevelt fell overboard from his yacht on the Bay of Fundy, a beautiful if freezing inlet between the eastern Canadian provinces of New Brunswick and Nova Scotia. The following day he was tormented by pain in his lower back and then, as the day progressed, he felt his legs grow increasingly weak until they could no longer sustain his body weight. This was the onset of Roosevelt's poliomyelitis, at this time known as 'infantile paralysis'. Poliomyelitis is caused by a virus that goes by the same name. In 1921 doctors were limited in their knowledge of the poliovirus, or indeed viruses as such. They might, however, have known that the virus did not infect Roosevelt while he was struggling in the cold water – the only infectious source of poliomyelitis virus is another person who has already contracted it. Once again, we are looking at an exclusively human reservoir. Moreover, the paralytic disease has an ancient pedigree.

Infantile paralysis was familiar to physicians in the time of the pharaohs of Egypt, since the effects of the disease were painted, with stunning accuracy, on the walls of their tombs. In 1921, as indeed today, there was no cure for the paralytic effects of the virus once it had afflicted a victim. Fortunately, Roosevelt was gifted with an extraordinary vitality and courage, enabling him to overcome the lifetime of paralysis that would result from his illness. It is to his credit that despite this handicap he became the 32nd President of the United States and he continued to serve the American people for an unprecedented four terms in office.

Viruses do not follow our human notions of rules and so they are apt to surprise us. One such surprise is that those viruses that replicate primarily in the gut – the so-called 'enteroviruses' – do not cause the usual symptoms of gastroenteritis. Instead, the viruses that do cause gastroenteritis are a miscellaneous group with members coming from widely different viral families. Of course, these include the genus of noroviruses within the family of caliciviruses. Another group of gastroenteritis-associated viruses are the rotaviruses, a genus within the family of reoviruses, which cause vomiting, diarrhoea and fever in babies under the age of two years. Other similar offenders include adenoviruses, coronaviruses and astroviruses. We are sometimes inclined to joke about the clinical effects of gastroenteritis, but the truth is that this is a distressing condition in people of any age. Moreover, in less developed countries, gastroenteritis is one of the commonest causes of death in children, a tragic situation complicating poor hygiene and contaminated water supplies. As we might anticipate, these illnesses are transmitted by the faecal-oral route.

The 'enteroviruses' are also transmitted by the faecal-oral route and the viruses also replicate within the intestine, but curiously they do not present with the typical fever, vomiting and diarrhoea that typifies gastroenteritis. Instead they cause less predictable and often complex patterns of illness that affect various organs and tissues, for example, the brain and meninges, or the heart, skeletal muscles, skin and mucous membranes, the pancreas, and so on. The most familiar of this strange gamut of enterovirus-linked illnesses is poliomyelitis. All three 'serotypes' of the poliovirus, which have slight differences in their capsid proteins, are 'enteroviruses' within the family known as the picornaviruses. We might recall that these belong to the family of very small RNA-based viruses that includes the rhinoviruses. A cardinal feature of enteroviruses is that they are resistant to acid, so they can pass through the human stomach to replicate further down the alimentary tract. The poliovirus was the first of the enteroviruses to be discovered, earning its finders – Enders, Weller and Robbins – a Nobel Prize in 1954.

We should not be too surprised to discover that humans are the exclusive host of the poliovirus. The individual virion is a mere 18 to 30 nanometres in diameter. Under the electron microscope it has a capsid with the familiar icosahedral symmetry, which encloses a relatively simple RNA-based genome. In the small intestine, the virus binds to a specific receptor molecule in the lymphoid tissues of the pharynx and the 'Peyer's patches' of the gut. Here the virus hacks its way into the interior of the cells, where it takes over the genetic processes to convert the cell into a factory for manufacturing daughter viruses. The daughter viruses are released through rupture of the infected cell, after which they re-invade neighbouring cells and repeat the process.

All of this sounds a trifle horrific and even potentially deadly. But in reality the great majority of individuals infected by poliovirus show little or no signs of disease other than, perhaps, a mild looseness of the bowels. But the stools of an infected individual will now be swarming with virus, which will be passed on to contacts through the faecal-oral route. Polio characteristically moves through populations in epidemic waves, with most of the infected unaware that they have encountered the virus. Only in a tiny minority does the virus make its way to the anterior horn nerve cells in the spinal cord, where infection and subsequent death of the nerve cells gives rise to the paralysis we saw in President Roosevelt. Bizarre as it might seem, the infection of the nerve cells appears to serve no purpose as far as virus transmission or evolutionary pathways are concerned. Indeed, this most dreaded complication of poliomyelitis appears to be coincidental.

The incubation period of poliovirus infection is usually a week to two weeks and, in the minority that show symptoms of infection, this involves a minor malaise, fever and a sore throat. These reflect the virus entering the bloodstream and will usually resolve without requiring any treatment and with no long-term consequences. Only in a small minority of those infected does polio give rise to a more severe illness. The onset is usually abrupt with headache, fever, vomiting – in some this may be accompanied by the neck stiffness typical of meningitis. Even still, the majority of symptomatic cases will go on to make a good recovery. But in the tiny but highly significant minority the paralysis of poliomyelitis sets in.

Paralytic poliomyelitis gets its name from the Greek *polios* for 'grey' and *muelos*, for marrow. This derives from the fact that the paralysis results from destruction of the grey marrow of the anterior horns of the spinal cord, which contain the cell bodies of the nerves that supply the muscles of arms, legs, chest and remainder of the trunk. The death of those cell bodies in the spinal cord causes a floppy style paralysis of the affected muscles, which is usually apparent within two or three days of the onset of the disease. In children affected by paralysis, this will have secondary long-term effects on limb growth and development. Bulbar poliomyelitis, a similar infection, causes damage to the nerve bodies of the cranial nerves, which results in paralysis of the pharynx and possibly accompanying difficulty with the muscles involved in breathing. This dreadful complication is why, before the advent of vaccination, some unfortunate patients ended up having to be supported by 'iron lungs'.

We do not know why this unfortunate minority of infected individuals develop serious disease, including paralysis, from the poliovirus. There is some evidence that the virus gets into the central nervous system more commonly than is suggested by clinical signs. Indeed, as we shall see, this pattern of unwanted penetration into the central nervous system can feature in illnesses caused by other enteroviruses. One wonders if some genetic propensity might perhaps play some role, but it may be no more than bad luck. As we saw above, this pattern of paralysis in children, with its effects on limb growth, was recognised in the wall paintings of the tombs of pharaohs from Ancient Egypt. How puzzling then that such an ancient and easily recognisable disease was unfamiliar to European doctors until the latter years of the nineteenth century, when the first epidemics began in the cooler climates of industrialised Europe and the United States!

Such has been the dramatic success of vaccination programmes, using live attenuated viral vaccines taken by mouth, that polio has been largely eliminated from developed countries. In 2018, according to the Global Polio Eradication Initiative, the disease is now endemic in just three countries: Afghanistan, Nigeria and Pakistan. But, given the ease and extent of modern travel, we cannot rest assured until this historic and maiming disease is completely eradicated in these remaining pockets of potential contagion.

While poliomyelitis is now approaching global control, it is not the only enterovirus to afflict humanity. Other members of this virus family are still commonly encountered in developed countries, including viruses that can be baffling in their presentations and clinically unpredictable in the course of their illnesses. Perhaps the best known of these are the Coxsackie B viruses, which sometimes present with a condition known to doctors as epidemic pleurodynia. Also known as 'Bornholm disease', after the Danish island where it was first recognised, this can present as severe chest pain arising from inflammation in the intercostal muscles of the chest wall. Popularly known as 'the devil's grip', the sudden onset and severity of the pain can mimic a heart attack. Coxsackie B viruses can occasionally cause inflammation of the brain, presenting as the condition known as myalgic encephalomyelitis, or 'Royal Free disease', named after the London teaching hospital where it first presented. The same enterovirus may also present with inflammation of the heart muscle, or myocarditis, coupled with inflammation of the membrane surrounding the heart, known as pericarditis, a combination that presents in both children and adults and can very occasionally prove fatal. Other enteroviruses, including the echoviruses and types 70 and 71 enteroviruses, can cause chest infections and various patterns of muscle, meningeal and brain infections, where the diagnosis of the causative virus may be exceedingly difficult to pin down.

Viruses and their associated illnesses can be very puzzling. Ever since we first discovered their enigmatic presence among us, questions have inevitably arisen as to the evolutionary purpose behind their behaviours. When faced with the unpleasant, sometimes life-threatening, effects of virus infections, we are inclined to wonder what possible benefit such behaviour might confer on the virus. In the case of the poliovirus we saw how it appears to be mere happenstance that the virus causes serious illness in a tiny minority of those it infects. But there are other viruses that sweep through the human population and inflict dreadful patterns of illnesses in the majority of those infected, sometimes accompanied by a high mortality. This is all the more baffling since all that matters to the virus is its survival and successful replication. Survival of the virus must surely be threatened by killing its host. When one views the same question from a medical perspective, we are inclined to question: why are some viruses so deadly?

7

Deadly Viruses

The Four Horsemen of the Apocalypse feature in the biblical Book of Revelation, where, having been released by the opening of seven seals, they ride out on red, white, black and pale horses. Theologians differ in their interpretations of what these riders might signify, but one of the four is commonly interpreted as pestilence, which, in modern terminology, would be interpreted as plague. While the common childhood infections, caused by viruses, are usually self-limiting, some viruses are truly dreadful in their capacity for death and suffering. In the recorded pages of history, two plagues of humanity would justify the term 'apocalyptic': these are the bacterial pandemics known as bubonic plague, as seen in the Black Death in the Middle Ages, and its viral counterpart, the plague of smallpox. Both have tormented humanity from ancient times, bequeathing a grim legacy in historical records and grave pits.

The Black Death was named after the festering swellings, or 'buboes', where lymph glands in the groin or armpit became swollen with pus and erupted onto the skin of victims. The causative bacterium, *Pasturella pestis*, is transmitted by the bite of an infected rat flea. Although the public commonly assumes that bubonic plague has gone away, in fact a milder form of the illness is still endemic in rural parts of the United States, South America, Asia and Africa. The viral apocalypse, smallpox, was named after the rash that accompanied the disease, which resulted from pustular blistering in the skin that healed with deep circular scars, or 'pocks'.

It is comforting to use the past tense here since, mercifully, smallpox has been eradicated as a plague. The clinical term for smallpox was 'variola', and the disease followed two very different patterns of virulence, depending on the causative virus. *Variola major* and *Variola minor* are species within the family of poxviruses. The poxviruses infect a wide variety of animals, but only three species infect humans: namely the two variola viruses and a related species, *Molluscum contagiosum*, which causes minor blisters on the skin of children. We shall confine our attentions to the variola viruses, which have a number of unusual features.

Humans are the only hosts for smallpox, so we are the exclusive reservoir of the two variola viruses in nature. The individual 'brick-shaped' virions are relatively large, measuring 302 to 350 by 244 to 270 nanometres. Before being displaced by the discovery of the 'Megaviruses', poxviruses were the giants among the viruses, being big enough to be seen as tiny cytoplasmic inclusions under high magnification of the light microscope. This feature alone alerts us to the fact that we are dealing with a relatively complex virus. The variola genome is predictably large and DNA-based. Unusually for a virus, it contains the genetic wherewithal for the manufacture of its own virus messenger, RNA, which takes care of the manufacture of viral proteins. This virus also has its own coded enzymes and transcriptional factors which control the manufacture of daughter viruses within the cytoplasm of infected host cells.

Smallpox viruses are extremely contagious, spreading by that most infectious route of all, aerosol inhalation. The viruses are also capable of spread through skin contact with the blistering rash, or through contaminated clothing, bed linen, utensils or dust. Infection usually begins with the arrival of the virus into the air passages of the throat and lungs of a susceptible individual, where they penetrate the superficial lining cells to be 'discovered' by the tissue macrophages, the first line of the human immuno-logical defences. The stage of infection within the macrophages is asymptomatic, but accompanied by stealthy advance of the virus towards its ultimate goal. By about the third day after infection, the 'virus-factories' within the macrophages journey on to the lymphatic stream and local lymph glands, from where the viruses spread to the other key elements of the 'reticuloendothelial system', in particular the bone marrow, spleen and circulating blood. This triggers a massive immune counter-attack on the virus, including cytotoxic T-cells and interferons. But, as the history, and the grave pits, suggests, this counter-attack is unsuccessful in the majority of sufferers. Symptoms begin with a severe sore throat at much the same time that blood-borne spread carries the viruses to the skin, where they produce the blistering and scarring rash, with its predilection for the face and limbs. The blisters are the result of direct viral invasion of the skin and they teem with viruses.

Historically it is thought that smallpox first arrived among humans about 10,000 years ago in the agricultural settlements in northeast Africa, spreading to India through trade with Ancient Egypt. It grieves one to imagine such a disease spreading through such populations of naïve people, and impossible to imagine exactly what they thought was among them. No doubt they had some simple rules for dealing with contagion, and, equally likely, they would have blamed some occult cause. We discover the pathognomonic pocks in the mummified skin of Ancient Egyptian mummies, such as the Pharaoh Rameses V, who died in 1156 BCE.

Smallpox, or the 'small pocks', was a clinical term that came into usage in the sixteenth and seventeenth centuries to differentiate it from the inch-or-more-diameter 'great pocks' that medical historians assume were pathognomonic of tertiary syphilis, a bacterial plague that may have been imported into Europe from the Americas. The viral plague of smallpox arrived into Europe much earlier, sometime between the fifth and seventh centuries CE, where it persisted as an infection, giving rise to repeated epidemics during the Middle Ages. Estimates suggest that it killed some 400,000 Europeans annually in the late 1700s, affecting all levels of society, including five reigning monarchs, and was responsible for a third of all cases of blindness. The same plague played a key role in the Conquistador subjugation of the Aztecs and Incas of South America, during the sixteenth and seventeenth centuries, when it may have dominated the history of encounters between Eurasian adventurers and the stricken native and hitherto 'virgin' peoples.

Today we can scarcely imagine the terror of living through a major epidemic of plague or smallpox sweeping through such a 'virgin' population. They would have been very quickly aware that a pestilence was among them, with panic-stricken populations in the grip of raging fever and, in case of smallpox, a virulent rash, which, when severe, caused the entire skin to boil with blisters and carried with it a horrific lethality, at its worst as high as 90 per cent. It must surely have seemed as if a pitiless demon had entered their world, intent on wiping out entire families, and even entire villages, towns and cities.

But smallpox was never a uniform death sentence. We cannot be certain of the actual levels of lethality of smallpox in various parts of the Americas, though we are informed that it was as high as 60 to 90 per cent in the worst-affected populations, falling to 30 to 35 per cent in some of the lesser-affected regions. This lower lethality was in fact similar with the calculated overall mortality of *Variola major* in concurrent Eurasian populations, suggesting that the virus had already become endemic in those regions. Meanwhile, even in the Americas, the *Variola minor* virus caused a much milder disease, with a mortality of about 1 per cent. It is somewhat ironic that smallpox, one of the deadliest plagues in history, was the first to be subdued by the use of a vaccine. Many readers will be familiar with the discovery of cowpox vaccine by the English physician, Edward Jenner, and this more than a century before the world even realised the existence of viruses.

In such less-enlightened times various therapies that we would now dismiss as 'quack' were touted as preventions or curatives for every frightening illness. In seventeenth-century England, for example, Dr Sydenham, an eminent physician in his day, treated patients in the throes of smallpox by allowing no fire in the room, leaving the windows permanently open, drawing the bedclothes no higher than the patient's waist and administering 'twelve bottles of small beer every twenty-four hours'. If nothing else, the beer would have dampened consciousness of the suffering – and perhaps the discomfort of the therapeutically induced hypothermia in winter. But it was famously known from ancient times that survivors of smallpox were immune to further infection. A hazardous treatment, involving inoculation of non-immune individuals with a scalpel wet with material from the ripe pustule of an infected patient, was variously employed in Africa, India and China long before Jenner introduced his vaccine.

History has it that Jenner overheard a dairymaid say, 'I shall never have smallpox for I have had cowpox.' Cowpox, a milder pox infection in cattle, was known as *vaccinia,* after the Latin, *vacca,* for cow. In 1796 Jenner conducted a now-famous experiment in which he inoculated an eight-year-old boy with pus from a vaccinia blister, obtained from a dairymaid with cowpox, and, having waited for the boy to develop immunity, subsequently tested this by inoculating him with smallpox. Thank goodness that the boy now proved to be immune. Although Jenner had rivals, who dismissed the importance of his discovery, the cowpox inoculation was soon taken up as a preventive measure against smallpox. We still refer to it today with the term Jenner coined for it: 'vaccination'.

When I was a child, it was still mandatory to be vaccinated against smallpox. I still bear the scar, which is a pock-shaped irregular oval about half an inch in diameter, on the skin of my upper left arm. Today children are no longer vaccinated against smallpox because the disease was eradicated from the global human population by a ten-year international programme of smallpox vaccination, headed by the American physician, Donald Ainslie Henderson, who worked under the auspices of the World Health Organization. This was formally signed off with the confirmed eradication of the disease in 1979.

There can be no denying that the eradication of smallpox was an extraordinary achievement. Ironically, however, this very success makes our modern populations unduly susceptible to a malicious attack involving a potentially bioengineered smallpox virus that might be deliberately created to be as lethal as possible. New generations, who have never been vaccinated, would have no inbuilt protection to such a spreading lethal strain. This is why the smallpox virus is now included in the list of Category A bio-warfare agents. Following smallpox eradication, it was agreed by international treaty that samples of the smallpox virus should only be retained in two maximum security laboratories – one at the CDC in Atlanta, in the United States, and one at similar facilities in Moscow, in Russia. The plan was to allow some continuing research aimed at countering any attempt to use the virus for bio-warfare, whether through terrorism or through formal hostilities between nations. We must hope that, if the worst comes to the worst, the officially sanctioned research in this small number of biosafety laboratories will come to our rescue with a modern vaccine, which will need to be spread globally with more efficiency than we have ever seen with any previous vaccination programme.

Why then are some viruses so lethal when they infect us?

We humans are graced with the gift of knowledge, education, morality, self-awareness, enabling us to think ahead, and so largely control the many aspects of our existence. Viruses are devoid of such self-knowledge, morality or anticipation. They are exclusively driven by those familiar goals: survival and reproduction. But it would be a mistake to underestimate them; viruses are extremely efficient in achieving those goals. Surely the lethality of dangerous viruses must be linked in part to whatever mechanisms viruses have evolved in order to overcome our human immune defences to viral infection. It will come as no surprise that the study of the smallpox virus in one of the two key bio-defence centres allowed to store the virus, the Centers for Disease Control in Atlanta, Georgia, has provided an important clue as to how the smallpox virus, *Variola major*, overwhelms our human immunity.

When *Variola* enters our human tissues, the 'innate immune response' is the first line in our defence against this alien invasion. As part of this innate response, the infected cells produce type I interferons in response to the presence of the virus, and these then engage other immunological defences that would normally inactivate and destroy the virus. What the scientists at CDC discovered is that inside the infected human cells the invading virus produces a protein, known as type I interferon-binding protein, which inactivates the human type I interferons. As we saw with the production of a virulence factor in norovirus infection, this is another example of the same malevolent viral strategy. In other words, the smallpox virus carries in its genome the coding for a major virulence factor that explains the severity of *Variola major* smallpox infection. This discovery of these interferon-binding proteins may, in the future, help authorities to improve the design of new vaccines and antiviral therapies, which might, for example, also apply to related viruses, such as the monkeypox virus, which causes virulent infections in humans.

In clinical terminology, 'virulence' is a measure of the severity of any viral, or other, infectious disease in any given host. In the case of viruses, this is the outcome of the interplay between viral infectivity and host resistance and susceptibility – what the old-time physicians used to call the battle between the soil, namely us, and the seed, namely the microbe. We now see how the production of these virulence factors is an important component of the lethality of smallpox. Indeed, we may infer from smallpox and norovirus that virulence factors may be common to many different viral infections. But we cannot simply assume this; we need to examine each specific virus and its relationship with its human host in detail before making any broad assumptions, because one of the things we have come to realise about viruses is that each virus is different in the ways in which it interacts with its host.

Clinical virulence can be measured in the starkest possible terms as the likelihood of death from infection from a specific virus. This is the parameter used in measuring the effectiveness of vaccines. Many vaccines, for example the MMR triple vaccine, work through 'attenuating' the virulence of a virus to the extent that it can be administered to large numbers of 'virgin' humans without causing clinical symptoms or signs of infection. At the same time, the vaccine induces resistance to a more virulent strain of the same virus, thus reducing the likelihood of serious illness, or indeed the risk of death, that the vaccinated individuals might otherwise encounter during the course of their lives.

Virulence factors are clearly important in virus infections, but they are not the exclusive explanation of the complexity of virus–host interactions across the diverse range of biodiversity. For a more comprehensive understanding of this we need to take into account the evolutionary circumstances of those same interactions between viruses and their hosts out there in the 'red-in-tooth-and-claw' of nature.

8

An All-American Plague

Let us remind ourselves that viruses are not sentient – they do not think ahead and plan a course of action. They have no morality: they are quintessentially amoral. They do not feel, hear or see . . . and only in the barest sense of recognition of host cellular receptors being presented to their capsid surface do they have the most primal sense of something akin to touch or taste. What drives them is also primal: survival in the battle to gain entry to the host's environment, through discovering some means of entry through the respiratory, alimentary or genital entrances – or through taking advantage of some external agency, such as a biting insect, to penetrate the tough protective hide of an animal or the covering epidermis of a plant – or that of a human. Once entry has been gained to the host's internal environment, the virus must survive the attack of the host's manifold immune defences, meanwhile searching for – or allowing itself to be found by – its target cell. Once discovered, the target cell (and sometimes there is a variety of target cells) becomes the natural ecology of the virus.

Within the cytoplasm, or perhaps the nucleus, of this target cell, the hitherto nascent virus particle becomes biologically hyperactive, discarding its protective capsid and laying bare its genome, with its quintessential predatorial genes, genetic sequences and proteins, to begin a fierce interaction with the host's genetic and related physiological and biochemical pathways. Here it is appropriate to remind ourselves that it is the virus's capacity to interact with this most intimate core of its host's being of that makes viruses so powerful, whether as agents of disease, or as agents of symbiogenetic evolution. Here in its procreative life cycle, the virus takes control of the host's genetic and related chemistry to make replicas of itself, these to be released as hordes of daughter viruses to continue the cycle of invasion and replication. This is the sole purpose of a virus, which surely encapsulates at primal level the purpose of every living being on Earth: to battle tooth-and-nail for its own survival, and to reproduce itself in the cruel unfeeling arena of nature. This was my perspective on viruses when I visited the scene of a dangerous emerging infection in the southwest of the United States in 1994.

A year earlier, a hitherto unknown virus emerged in the Four Corners states of New Mexico, Arizona, Utah and Colorado, causing widespread local panic. The index cases were on the Navajo Nation reservation but it soon became apparent that the emerging virus had no particular association with the Navajos, but rather with communities living in rural locations. Within six weeks of the outbreak, the causative virus was found to be a hantavirus, diagnosed by molecular geneticists working at the Centers for Disease Control in Atlanta. Indeed, once the virus was isolated and subsequently cultured, it was seen to be a new species of hantavirus, which acquired the name *Sin Nombre* – the virus without a name. Hantaviruses are species of bunyaviruses, an order of RNA-based viruses that include some very nasty pathogens of humans, including California encephalitis virus, Rift Valley fever virus, Oropouche virus, haemolytic fever with renal syndrome virus and Crimean-Congo haemor-rhagic fever virus.

I visited the Four Corners area as part of my researches for a book, *Virus X*, in which I set out to examine the circumstances and evolutionary behaviour of emerging viruses. I was looking for answers to some troublesome questions. 'Where do emerging viruses come from? Just how dangerous are they? Why do they behave with such lethal aggression? What can we do to protect ourselves from them?'

The *Sin Nombre* hantavirus epidemic was still ongoing locally, and it had proved to be very aggressive indeed. The local consultants in intensive therapy and respiratory medicine in the teaching hospital in Albuquerque were kind enough to allow me to interview them, and they also allowed me to speak to some of their affected patients on the intensive care units, recovery wards and follow-up outpatient clinics. It was an enlightening experience and I remain indebted to those colleagues and to those patients and families. I shall call one such patient Marianne, and her mother, Joanne. When I met Marianne, she was a slim woman, 21 years old with blonde hair cut boyishly short. She was wearing faded jeans and a matching denim jacket over a blue t-shirt. A reflection of her recent brush with death still showed in her eyes, and in a jerky nervousness in how she moved her body. Asked a question, there was a slight hesitancy before replying, but then she spoke quickly, a scattering of words expressed in the delightful Western accent typical of New Mexico:

'This has kinda taught me a lot,' she said a little shyly. 'So I'd like my story to help other people.'

Joanne and Marianne lived in one of the small towns off the famous Route 66, which passes through the Four Corners states. Two months earlier, on 23 May, Marianne had spiked a fever.

Joanne, a qualified nurse, thought it was no more than a touch of flu and she treated Marianne with Tylenol and aspirin. Later in the day, Marianne felt nauseated and rested on the couch. That was how it continued over the next two days. Then the fever worsened and Marianne developed a severe aching in her muscles. 'It was a real dull sort of pain, in my shoulders, my thighs, my calves, in my back. Every time I moved it hurt.' The following day Joanne went to work in a nursing home, 60 miles up the road. When she called home to enquire about Marianne, what she heard alarmed her. Marianne could hardly talk because of the difficulty in her breathing. There was a rattling in her throat and her temperature had soared to 39°C. An alarmed Joanne called Marianne's grandmother and asked her to take her into the local hospital.

There was a nagging fear in Joanne's mind as she hurried back home, down Interstate-40. 'Oh God – I hope Marianne hasn't got the hantavirus!'

When she arrived at the hospital Joanne was shocked by Marianne's appearance. Her daughter was extremely breathless. Her lips were purple and her nail beds were blue. A circle had appeared round her mouth that was a livid purple and her skin had turned grey, the colour of slate. Her mother helped her make repeated trips to the toilet, where she had copious bouts of vomiting and diarrhoea. She told the hospital staff of her fears that Marianne had the hantavirus, but they didn't believe her. To be fair to them, this scepticism wasn't altogether surprising; the hantavirus was provoking local panic, but in most local hospitals it was a relatively rare diagnosis. Instead the staff were convinced that Marianne had common-or-garden gastroenteritis. So they treated her for this, pouring intravenous fluids into her veins to correct her dehydration.

But Joanne could see that this was not curing her daughter. She was convinced that Marianne was dying right there before her eyes.

In a growing desperation, this nurse and mother found herself in the position of fighting a desperate personal battle to save her own daughter's life. She became hysterical. She yelled and screamed at the staff and then called in her family doctor, who took one look at Marianne before calling the air-ambulance helicopter to have her airlifted to the University Hospital in Albuquerque. During the 70-mile journey Marianne kept blacking out. When I spoke to the nurse who first took care of Marianne on her arrival at the hospital, she recalled that Marianne cried out: 'I'm drowning – I'm drowning!' The *Sin Nombre* hantavirus gives rise to a mixture of cardiac and pulmonary failure, hence the syndrome 'cardiopulmonary syndrome', and it causes the lungs to fill up with fluid so that afflicted patients literally drown in their own pulmonary secretions. What Marianne was describing was exactly what was happening inside her lungs.

But now the struggle for Marianne's life had passed over to a dedicated team of doctors and nurses in the intensive therapy unit who had been fighting the same desperate battle against the *Sin Nombre* hantavirus over the preceding year. Marianne's chest x-rays showed that her lungs were a complete whiteout. Her breathing was now taken over by a ventilator. Her cardiac trace was showing electrical irregularities. Over that day and succeeding days, she would suffer from serious disturbances in her heart rhythm. The doctors talked about putting her onto the ECMO machine, which is similar to the heart-lung bypass machine used in open-heart surgery. Joanne signed the necessary papers in case this should prove necessary. Hour by hour it remained touch and go. Several of Marianne's internal organs began to fail. She developed pancreatitis – inflammation of the main digestive gland in the abdomen – that was life-threatening in itself. Her liver began to fail. Her bone marrow became depressed, causing anaemia. Her blood pressure would swing about crazily, leaping to dangerous elevations or falling to very low levels. Every complication required further urgent treatment.

For four days Joanne refused to leave her daughter's bedside. She would not leave her even to sleep. 'I would stand there watching the monitors – I knew enough to read them but too little to be able to do anything about it. I felt so utterly helpless. I remember seeing the cardiogram with premature ventricular contractions, just wham, wham, wham, wham, one after another. Then I thought, that was it . . . We're not going to come out of that one, because the whole line was PVCs.' (PVCs are premature ventricular contractions.)

After the ordeal of those first four days, and thanks to the heroic determination and dedication of the medical and nursing staff on the intensive therapy unit, Marianne showed her first signs of improvement. She started clearing fluid in large quantities. Only now would her mother allow herself to go home and get some rest. Marianne would stay on the ventilator for two and a half weeks. Confused by her sedation, she needed to be strapped to the bed so the only direction she could look was up. Joanne pinned photographs of Marianne's son, then eleven months old, to the ceiling above her head so it would be the first thing Marianne saw when she woke. One day, at 1.30 a.m., she pulled her hand free from the restraints and she extubated herself there on the intensive care unit. After just a few more days this extraordinarily brave and physically resourceful young woman came home with her equally brave and resourceful mother, finally cured of the disease.

The *Sin Nombre* hantavirus spread to many other states beyond the Four Corners area before it gradually settled of its own accord. The source of the virus was found to be the commonest wild mouse in the US, known as the deer mouse, or *Peromyscus manic-ulatus*, in which the virus causes no apparent disease. Virologists describe the deer mouse as being the natural host of the virus. It also became apparent that humans, like Marianne, became infected with the virus when they came into accidental contact with deer mouse urine, saliva or faeces. This did not mean that families such as Joanne's were careless or unhygienic in their habits, it just meant that they lived in a rural environment where contact with wild mice was more likely.

Why, then, had the epidemic broken out in 1993?

Local biologist, and rodent expert, Bob Parmenter had been conducting ecological studies in the local habitat, and handling deer mice for years. In an interview with the local media, he declared: 'It's hard to believe that such an adorable little animal could cause so much trouble.' With tawny fur, big protruding ears, lustrous black eyes and an inquisitive rounded snout with prom-inent black whiskers, the mouse looked more like a friendly little personality from the tales of Beatrix Potter than a threat to people. But zoologists such as Parmenter knew that the deer mouse is exceptionally tough and enduring. He had witnessed a striking example of this when he was studying the recovery of the ecology in the area around Mount St Helens after the devastating volcanic eruption of 18 May 1980. At that time it had intrigued him to discover that the deer mouse was the first animal to recolonise the landscape. Deer mice are tough, obstinate creatures. Overcoming any privation and prepared to eat whatever is going, they never hibernate, produce as many as five litters a year, and can conceive while nursing the previous litter.

Parmenter was well aware that deer mice readily found their way into every nook and cranny of their environment, including human homes – and even into vehicle air ventilation systems. Over the months prior to the human epidemic, he had noticed massive increases in mouse populations, as much as thirty-fold in some areas. It seemed more than fortuitous that the same geographical area was the epicentre of both the explosion in the mouse population and the outbreak of the lethal virus. The two had to be causally related. Why then had the mouse population seen such a massive increase?

After seven years of drought, and perhaps because of an El Niño climatic effect, New Mexico had enjoyed two mild winters with more rain and snow than usual. Parmenter's data showed that this mild weather had nurtured a heavy crop of piñon nuts and insects such as grasshoppers, the favourite food of the mice. The increase in food sources had resulted in a burgeoning of the mouse population. The mere fact that humans and mice shared the same ecology would, very likely, have been enough to trigger the epidemic.

As of January 2017, 728 cases of hantavirus were reported in a series of relatively small outbreaks across 36 American states, mostly west of the Mississippi River. The medical term for this type of outbreak is 'endemic' as opposed to 'epidemic'. How very fortunate that the *Sin Nombre* hantavirus failed to jump species to establish a new host in humans. Scientists have been asking themselves the very important question: why not?

It is likely that there were multiple reasons. Rodents live in unhygienic burrows where their young would inevitably be contaminated by secretions and excreta, meanwhile modern humans are hygienic, with vacuumed homes and toilet waste disposal and hand-washing. Yet despite our modern human hygiene other viruses have readily spread to become epidemic amongst us. Perhaps we need to look a little deeper at what happened when humans, such as Marianne, became infected with the hantavirus.

We know that the incubation period is long. Early symptoms of hantavirus pulmonary syndrome present like the flu, with muscle aches, fever and fatigue. But unlike flu, with its very rapid development, the hantavirus symptoms only show up some two to three weeks after first contracting the virus. We also know that the virus proliferates in the lungs, spleen and gallbladder. The pulmonary effects show up some four to ten days after the onset of symptoms, and include the difficulty in breathing that we witnessed with Marianne. The subsequent focus of the 'hantavirus pulmonary syndrome' on the lungs, with the massive outpouring of oedema fluid that causes the patients to drown in their own secretions, would suggest that there might have been a potential for a spread from human to human through coughing. Had this actually been the case, it might have offered the terrifying potential for aerosol respiratory spread, the deadliest pattern of transmission, such as we see in influenza. The virus managed to get to the tiny blood vessels that come up close to the air-sacs of the human lungs as part of oxygenation of the blood. Mercifully it failed to cross over a membrane that was only a few cells thick. The world was lucky that year.

It is important to ask ourselves another question: why were we so lucky?

The first and most obvious reply would be that the virus has not evolved to infect and then become transmissible between humans. It is a rodent virus. Thankfully we are not the natural reservoir. The dramatic human infections, the illnesses and deaths, were caused by accidental transfer of the virus into a new and alien host, humans who had no previous experience of contact with the viruses.

During my time spent with the biologists, doctors and other scientists dealing with the hantavirus epidemic in America, I learned something new and very important about viruses – indeed, it proved to be a revelation that changed the course of my subsequent career. At this time my opinion of viruses would have been typical of most doctors: they were genetic parasites that caused nothing but disease. But back there, in 1994, when I interviewed another senior biologist involved in the hantavirus research, Terry Yates, Professor of Zoology at the University of New Mexico, he explained to me that every rodent has a species of hantavirus that is co-evolving with it. This came as something of a surprise to me.

What, I wanted to know, does this term 'co-evolution' mean?

He explained with a hypothetical example. We are familiar with the duck-billed platypus as an egg-laying marsupial, but let us pretend that it is a rodent and we are faced with the problem of placing it in its appropriate position on the evolutionary tree of the rodents.

I sat back even more mystified.

He explained that, if I were to provide him with the RNA-based genome of its hantavirus, he could place the virus with pinpoint accuracy into the phylogenetic tree of the hantaviruses. And then if he were to overlap the two phylogenetic trees, that of the hantaviruses and that of the rodents, he could now identify the exact position that the platypus occupied on the phylogenetic tree of the rodents.

'The two trees exactly parallel one another?'

'Yes.'

'How could that happen?'

'They are co-evolving with one another.'

This appeared to imply an extraordinary intimacy between the evolutionary history of the hantavirus and the evolutionary history of the rodents. I thought about this for a while. That interview with Professor Yates, which had been scheduled to last an hour in his office in the biological museum in Albuquerque, now stretched into several days, during which he courteously put me up in his home and introduced me to his family and colleagues. I accompanied him on a trip to the nature reserve called the Sevilleta, where Professor Yates, Bob Parmenter and generations of earlier zoologists had been studying deer mice for more than a century, amassing huge collections of specimens whose tiny bodies were laid out in a vast array of trays in the university museum. Those tiny bodies were now available to the biologists for the study of the companion viruses.

During this time, and amid the many conversations I had with Terry Yates and his colleagues, I learned a great deal more about the extraordinary co-evolutionary link between viruses and their rodent partners. This caused me to ask what I now believed to be a critical question:

'Could it be that hantaviruses and rodents are each influencing the partner's evolution? If so, surely this has to imply that they are involved in a symbiotic relationship?'

He looked at me and shrugged.

We know that viruses do not think. Viral behaviour is controlled by evolutionary forces. It now seemed to me that what Terry Yates referred to as co-evolution of virus and host must imply a symbio-genetic pattern of evolution between hantaviruses and their host rodents. After I got home I researched what I could find of the literature on viral symbiosis. I came up with very little. I found no mention of d'Herelle in what searches were available to me at that time. I did, however, discover that entomologists had used the term 'symbiosis' in relation to parasitic wasps and their viruses, which are called polydnaviruses. I would, some years later, discover that one or two colleagues had also raised the idea of symbiosis

81

in relation to retroviruses, the type of viruses that cause AIDS. But it seemed to me that the concept of viral symbiosis had never been formally defined and examined in the systematic way in which a scientific concept should be developed.

If there was little in the way of reference to symbiosis in relation to viruses, symbiosis was a familiar enough concept in general biology. I discovered that an American scientist based at Amherst, Professor Lynn Margulis, was an expert on symbiosis as an evolutionary dynamic. I also came across the name of the distinguished Nobel Laureate, Joshua Lederberg, who was currently President of the Rockefeller University, in New York, who had used the term in relation to bacteria and their phage viruses. I wrote him a letter and he kindly agreed to my interviewing him. During that interview, he confirmed that there were examples of viral symbiosis in the complex relationships between bacteria and their phage viruses. But when I asked if he had come across any examples of viral symbiosis in relation to plants and animals, he replied: 'I don't know of any examples – but I think it would be interesting to go look for them.' He also advised me to look up Lynn Margulis, whom he had once taught genetics.

I took his advice on both counts. I began to search for examples of viral symbiosis in the scientific literature. I also looked up Lynn Margulis, who graciously agreed to my interviewing her about her life and work with symbiosis. We ultimately became friends. I read many of her books and found her inspirational when it came to understanding symbiosis *per se* and symbiogenesis, which is the study of symbiosis as an evolutionary force. But Lynn was not knowledgeable about viruses. The results of my initial researches on emerging viruses, and my tentative explorations of the evolutionary role of viral symbiosis, were presented in the book *Virus X*, which was published in 1997. I developed a novel symbiological concept in which, as part of their evolutionary strategies, viruses actually contributed aggression, sometimes exceedingly lethal aggression, as a quintessential part of their symbiotic inter-action with their hosts.

At this stage it occurred to me that, given the involvement of viruses with the genetic landscape of their hosts, if viruses were to change the host genetic landscape in such a way as to be bene-ficial to the host, the evolutionary pressures on the host would ensure that this benefit would be selected for as part of the evolu-tionary development of the host. To further help me to extrapolate symbiosis and symbiogenesis to viruses, I subsequently made contact with another distinguished colleague, Professor Luis P. Villarreal, who is a globally-acknowledged expert on evolutionary virology. I interviewed Luis in some detail over the phone, when we discovered that, though we had approached the theme from different directions – in my case the symbiological perspective, and in Luis's case the classical Darwinian perspective – we had arrived at very similar conclusions. Just as Lynn Margulis had become my mentor in relation to symbiosis as an evolutionary force, Luis Villarreal became my mentor in opening my mind to the important role of viruses in the evolution of life. Key to his accepting my concept of viruses as symbionts was my recognition of the potential of viruses for 'aggressive symbiosis'.

I now came to realise that this aggressive pattern of the inter-action between the *Sin Nombre* hantavirus and humans had commonalities with other 'emerging' viruses, such as HIV, SARS, Ebola and avian influenza. In all of these examples the relevant virus, which behaved in a very aggressive way to humans, showed a striking lack of aggression in relation to its long-established zoonotic host. What made no sense when examined from the medical perspective made perfect sense when examined from a

non-human-orientated evolutionary perspective. This changed the
way I viewed viruses. It did not alter the fact that we, as doctors,
must look to stopping viruses in their tracks when it comes to
viral illness in our patients, but it highlighted the need for a broader
vision of the role that viruses might be playing out there in the
natural world.

9

Lurker Viruses

The medical name for the virus that causes 'cold sores' is herpes simplex. The Greek word, *herpeton*, refers to reptiles, such as snakes and lizards, which have creeping or crawling manners of locomotion. The creeping spread of the blisters around the mouth and genitals, and their connection with the sexual passions, would have been obvious to the Greeks of Hippocrates' day, leading those ancient physicians to compare the condition of herpes with reptilian motion. Shakespeare was also acquainted with the genital affliction, and its passion-driven transmission, as when, in *Romeo and Juliet*, he described the punitive actions of Queen Mab, midwife to the fairies: 'O'er ladies' lips, who straight on kisses dream, which oft the angry Mab with blisters plagues . . .'

The herpesviruses are a family of relatively large viruses, divided into three broad sub-families, in all comprising more than 130 different species, which infect mammals, birds, fish, reptiles, amphibians and molluscs. Like the *E. coli* bacterium, and indeed *Homo sapiens*, the herpesviruses have genomes comprised of double-stranded DNA. But the viral genome is much smaller than that of the bacterium, and vastly smaller than our own. The typical herpesvirus, when wrapped in its enclosing envelope, is from 120 to 200 nanometres in diameter. This is considerably larger than we saw with the picornaviruses. The sizeable viral genome is protected by an icosahedral capsid built up of 162 tubular capsomeres, with the whole further ensheathed in a baggy envelope made up of host lipids and viral proteins. Even so, the virus is still much smaller in size than a bacterium and it lacks the cellular properties of a bacterium.

A simple way of looking at such differences is to grasp that a human cell is vast when compared to a bacterial cell and a bacterial cell is vast when compared to most viruses. But when we look more closely at the genomes of all three organismal types, we make a curious discovery. Unlike the contiguous ring genome of the bacterium, our human genome is made up of 46 separate linear stretches of DNA, called chromosomes. Each chromosome is in fact a single, extraordinarily long, molecule. In another book, *The Mysterious World of the Human Genome*, I used the metaphor of 46 separate railway tracks for our chromosomes, tracks that an imaginary steam train could travel along from start to finish, to make some interesting journeys and stopping places.

How curious then to discover that, unlike the bacterial genome and rather more like our human chromosomes, the herpesvirus genome also consists of a single linear thread of double-stranded DNA, which is packed to fit within its capsid like a tight-wound spool of a fishing line. The viral genes distributed along this thread code for as many as 100 proteins, many of which are enzymes, including the viral DNA polymerase, which is essential for viral replication within the nucleus of its host cell. Doctors treating herpetic conditions take advantage of another of the virus's own enzymes, known as thymidine kinase, which is tricked into activating certain antiviral drugs.

Some nine species of herpesviruses are responsible for diseases in humans. The most familiar of these is the herpes simplex virus, or HSV, which causes those discombobulating cold sores and other more intimate rashes. Once again we discover that we are the reservoirs – the natural hosts – of these viruses, which have entered into an aggressive symbiotic relationship with us. Cold sores are caused by two closely related species, HSV-1 and HSV-2, either of which can give rise to body-wide manifestations. As a general rule, HSV-1 has a predilection for the upper parts of the body while HSV-2 has a predilection for the genitals, but such predilections are not an absolute guide. Unfortunately, there is little by the way of immunological cross-protection between the species, so that a victim of one of the species has no guarantee of avoiding future infection by the other one.

What then is going on in an individual afflicted with herpes simplex?

Within those stinging blisters of the mouth or genital areas, the first step in human infection involves the virus binding to and then fusing its surface membrane with the cell membrane of the skin or mucous membranes. This enables the virus to enter the cell interior, or cytoplasm, where it sheds its surface membranes and heads for the nucleus. Here, in the landscape of the human genome, many of the herpes viral genes come into play, notably the powerful viral DNA polymerase, which replicates the viral genome; meanwhile, other viral genes are transcribed to various 'messenger RNAs' that will give rise to viral structural proteins, such as those found in the capsid, so that in typical viral fashion the end result is to convert the cell nucleus, and its closely linked genetic and biochemical pathways, into a factory for daughter viruses. The end result is the death of infected cells, which rupture to release hosts of daughter viruses that go on to repeat the cycle by infecting other host cells.

'Primary infection' is what happens when an individual first becomes infected with HSV-1 or HSV-2. This tends to happen in infancy or early childhood from close contact with an older, previously infected person, often as the result of a kiss. Such primary infection will often produce little or no overt symptoms or signs. On rare occasions it results in an unpleasant illness with fever, accompanied by painful blistering on the lips and inside the gums and membranes of the mouth, the blisters breaking down to form shallow ulcers. These tend to crop in the front areas of the mouth as compared to the vesicles of a different virus, known as the coxsackievirus, which manifest further back on the hard palate and the back of the throat.

If we were to examine the fluid within the blisters, we would discover cells in the process of swelling up and detaching from the plasma membranes, while other cells are in the process of rupturing or becoming fused together into multinucleated giant cells. The body would already be responding with its armamentarium of countering defences, involving the quick-forming antibodies, known as the immunoglobulin IgM, followed by the more powerful and long-lasting IgG, aided and abetted by a mobile army of cellular immune defences. As this miniature war progresses, the blister fluid turns pustular, and as the defences win and the virus is eradicated, the blisters dry up to become scabs. Fortunately, scarring, such as is seen with smallpox, is uncommon but can occasionally happen in patients with frequent recurrences.

In most cases of primary herpesvirus infection, the fever and rash are self-limiting, with the body's immune system bringing things under control and leading to healing within two weeks or so of the onset.

Alas, there is currently no immunising vaccine for HSV-1 and 2. But antiviral drugs, for example Acyclovir, can be helpful in treating primary cases and secondary recurrences. In severe cases the drugs can be administered intravenously, but they are more commonly taken by mouth or administered topically in a gel or cream. Despite such therapy, the virus doesn't usually go away but rather lurks within the body, perhaps for life. This is why even in later years the blisters may break out again, usually around the mouth and heralded by a prodromal itching. With such recurrences, the watery blisters, which teem with viruses, crust over and heal within a few days. Thankfully these recurrences tend to become less frequent with the passage of time and may eventually stop altogether.

Genital herpes, affecting the skin of the labia, vulva and perineum in women or the penis in men, is, as Shakespeare's Queen Mab was spitefully aware, transmitted by means of sexual intercourse. The rash can also spread to the upper inner thighs and sometimes the cervix in women and it can involve the perianal skin in homosexual males. The regional lymph glands at the top of the thighs may be swollen and tender and the rash may be accompanied by fever and, more so in gay males, may be accompanied by viral meningitis. These days sexually transmitted infections are on the increase, in spite of our greater understanding of the infectious agents and the potential for prevention through the use of condoms. They are apt to provoke considerable anxiety, social upheaval and emotional distress. In this modern age, sufferers tend to find outlets in the social media, where they can search for useful ad-hoc advice, discover support groups and the comfort of shared experiences.

As with oral herpes, the viruses of genital herpes stick to character and hang around even after the infectious manifestations settle, and so the primary infection may be followed by recurrent infections, albeit usually less severe in their symptoms and manifestations. Why, we might well ask, do the viruses trouble us again and again, when they should have been recognised and done away with at first infection by our immune system?

To discover why we need to look more closely at what is really going on during that primary interaction between the viruses and their human hosts. Although symptomatic infection appears to be localised to the mouth or genital epithelia, in fact the viruses concomitantly invade the local lymph glands and, as is the way with so many viruses, find their way into the bloodstream. It is during this bloodstream, or 'viraemic', spread that – albeit only in rare cases – the HSVs can cause meningitis, or even rarer still, the inflammation of the brain called 'encephalitis'. Such serious complications are more likely to trouble people who are immunocompromised. In such cases intensive in-patient therapy will usually be warranted.

But we have not yet addressed the reason why the condition comes back after apparent initial healing!

During the phase of deeper and usually silent spread, the viruses find their way to the local sensory nerves, using the nerves as a conduit to get to the local 'nerve root ganglia', akin to nerve distribution centres. It is here, in some manner that is currently not fully understood, that HSVs become latent, lurking for what may be many years until some unknown stimulus, perhaps the local skin inflammation that follows that daft holiday sunburn, or perhaps even a physical or mental shock – anything that might temporarily weaken our local immune defences. Somehow this acts as a stimulus for the reactivation of the viruses, and down they come along those same regional nerves to invade the skin and provoke those creepy-crawly blisters to start cropping over our mouths and private parts again.

★

When is a 'pox' infection not really caused by a 'pox' virus? The answer is when it is the familiar childhood infection that we call chickenpox. In fact, the label is doubly misleading, since not only is the condition not caused by a pox virus, it also has nothing whatsoever to do with chickens. Whatever the misleading name of the condition, the rash of chickenpox, which might have understandably been confused with milder infections with smallpox in the past, is caused by another species of herpesvirus, known as the varicella-zoster virus, or VZV.

As with all of the viruses within this chapter, the varicella-zoster virus is exclusive to humans. The word *varicella* derives from the Latin for a smallpox-like pock and it refers, of course, to the chickenpox rash. VZV is transmitted by the highly contagious respiratory route of coughing and inhalation. As the name varicella-zoster implies, this same virus causes two very different patterns of illnesses: the blistering febrile rash, which usually presents in children, and the painful, sometimes excruciatingly so, rash of shingles, or 'herpes zoster', in adults.

Chickenpox announces its arrival with the characteristic rash, which takes the form of flat red spots that evolve to bumps and then blisters, more florid on the face and body than the limbs. There is a mild accompanying fever and the rash can pop up again in recrudescent crops before the blisters eventually subside to crusts that are shed as part of the process of healing. In infrequent cases the blisters can become secondarily infected with bacteria and, more rarely still – for example in immunocompromised patients, such as children suffering from leukaemia – the virus can cause a life-threatening pneumonitis or encephalitis. Fortunately, the vast majority of chickenpox sufferers will make a full recovery without any scarring.

Being a herpesvirus, VZV shares the lurker tendency with its family, hence its ability to disappear for decades and then to make an unwelcome reappearance in a different guise. This is where the 'zoster' half of the name comes from. As we saw with HSV, varicella-zoster also lurks in the nerve ganglions. But unlike HSV, which tends to be restricted to the ganglions of the facial and genital sensory nerves, VZV spreads via the bloodstream to discover the many different sensory nerve ganglia throughout the body. This is why it can manifest later in life, perhaps at a time when the individual's resistance is low, as a painful blistering rash in unpredictable skin distributions, whether on the face, the chest or abdomen. On the trunk it tends to sweep around in a girdle-type distribution following the sensory nerve girdle-like 'dermatomes'. This is actually where the name zoster comes from: *zoster* is the Latin name for girdle.

From this we gather that people who suffer from shingles must have suffered a preceding chickenpox virus infection. Indeed, those suffering from shingles should beware the fact that the blisters contain infectious viruses and so they should avoid contagion to children, or even adults, who have no prior history of chickenpox infection.

*

Thus far we have considered two of the commonest and most familiar of the herpesviruses, which, despite their lurker tendencies, follow predictable histories in their infectious behaviour. Other members of the same virus family are less predictable. These include the cytomegalovirus, or CMV, which, though it is actually one of the commonest viruses to afflict people in the Western world, may be less familiar to my readers. The name, cytomegalovirus, implies that this virus causes cells to swell up, with large unnatural inclusions within the infected cell nucleus. A problematic facet of the virus's presentation is its very unpredictability, since it produces different patterns of presentation at different ages – and it is manifestly capable of infecting people of all ages from the cradle to old age.

Clinical manifestations of CMV are relatively rare. But if a pregnant mother is infected with CMV, she may inadvertently transmit the virus to the foetus via the placenta, resulting in severe sickness in the newborn or even the tragedy of fatality. Babies can also pick up the virus through breastfeeding, when it seems that the antibodies in breast milk provide no adequate protection. Stranger still, infection during infancy, or subsequent childhood, may also result in no symptoms or signs of illness. But this does not mean that the virus has been eradicated from the body – rather it may be exhibiting that characteristic lurker tendency of the herpesviruses. By the time the child reaches adolescence, the latent infection may become overt, giving rise to lassitude, fever, impaired liver function – indeed, it may present as 'infectious mononucleosis', which is more commonly associated with yet another herpesvirus, called the Epstein-Barr virus, or EBV. Both CMV and EBV are capable of being transmitted by kissing and sexual intercourse. Infectious mononucleosis, also known as glandular fever, can result in abnormal-looking lymphocytes in the peripheral blood smear and the illness, which commonly manifests in young adults, is often accompanied by enlargement of the spleen.

Infection with the cytomegalovirus is far commoner in Western countries than many of us realise. Astonishingly, between 50 per cent and 80 per cent of all Americans are reportedly infected with CMV by the time they reach 40 years old. In typical lurker virus fashion, once acquired, the virus never completely goes away. Equally remarkable is the fact that the majority of such CMV 'carriers' show no symptoms whatsoever of illness from the virus. In fact, if we put this observation together with the behaviours of HSV and VZV above, this capacity to lose themselves within the ecology of the human body appears to be a shared common characteristic. But it would be a mistake to assume that the presence of CMV is always benign. As with the other herpesviruses, CMV can produce serious disease if the immunity of the host is compromised, for example, in the very young and very old, and in people where some other illness, or perhaps medical onco-therapy, causes significant immunosuppression.

We should, perhaps, recall the fact that viruses are capable of surprising us. And a virus that inhabits so very many people – for the most part provoking little or no signs of illness – might suggest a potential symbiotic partner. Indeed, there is some evidence that CMV can sometimes prove beneficial to its human host. One of the tissues in which CMV lies dormant is the myeloid portion of the bone marrow. These myeloid cells play a key role in our normal immune defences. There is some evidence that the presence of CMV in these cells may actually improve our immune response to other infectious agents that have invaded our body and entered our bloodstream. Such a protective role for 'endogenous viruses' has also been reported in relation to other groups of viruses, including retroviruses, which we shall come to examine later in this book. It would appear that sometimes lurkers might, in reality, be more loiterers with mutualistic potential!

This might be an appropriate time to become better acquainted with a term that evolutionary virologists have introduced in relation to how we should regard viruses: that term, and concept, is the so-called 'virosphere'.

This perspective suggests that life, from its very beginning, has existed within, and been deeply influenced by, an invisible enveloping ecosphere of viruses. The concept is relatively new and so radically different from former perspectives of viruses that it will naturally provoke scepticism in some quarters. Nevertheless, it is supported by evidence coming from new ecologically orientated studies called viral metagenomics. This is one of the most rapidly expanding fields of study at present, and one we shall explore in more detail in subsequent chapters. We should remind ourselves that the definition of viruses as symbionts needs to be qualified by adjective 'aggressive'. Also, for the moment, perhaps we might give a thought or two to considering the rather strange observed behaviour of herpesviruses as permanent inhabitants of the human ecology in relation to this concept of virosphere and its broader consideration of viruses in relation to life on Earth.

The Epstein-Barr virus also happens to be a herpesvirus, but it has rather more sinister properties than the other members of the same viral family. In 1958 an Irish surgeon, Denis Parsons Burkitt, wrote a paper about a malignant tumour in African children that was prevalent in regions with a high incidence of malaria. A few years later, Burkitt was delivering a lecture on the condition at a London hospital, during which he showed slides of afflicted individuals with grossly swollen jaws infiltrated by tumours. Remarking how the tumour was found in regions where malaria was endemic, Burkitt wondered if the tumour might be caused by a virus spread by the mosquito. Sitting in his audience was a pathologist, called Michael Anthony Epstein, who had a keen interest in the electron microscope.

Epstein, with colleagues Bert Achong and Yvonne Barr, subsequently proved that the cancers were indeed caused by a virus, not a mosquito-borne virus but a newly discovered herpesvirus. Today we know it as the Epstein-Barr virus, or EBV. Cell lines containing the virus were sent to Werner and Gertrude Henle at the Children's Hospital of Philadelphia, who developed serological markers that recognised the presence of the virus in an infected individual. In 1967 a technician in their lab developed glandular fever, with the typical mononucleosis blood findings, and the Henles were able to confirm that the virus that caused cancer in African children also caused glandular fever in their technician. The following year they took their lab investigations further to discover something remarkable. When the immune cells, known as B-lymphocytes, were infected with the Epstein-Barr virus, the cells became immortalised. This was a pioneering discovery of yet another mysterious capacity of viruses: they can change the developmental fate of cells. Cancer cells, in their peculiar way, are similarly immortalised.

Even today we do not fully understand EBV. But we do know a great deal more about it. We know that it is a species of human herpesvirus, called HHV-4. We also know that it is one of the commonest viruses to 'infect' humans. I have placed the word in quotes because discoveries such as these really do question what we really mean by the nature of virus interactions with their hosts. Today we know that EBV is the commonest cause of infectious mononucleosis, also known as glandular fever. It is also the cause of Burkitt's lymphoma, and it may be somehow associated with some cases of Hodgkin's lymphoma, gastric cancer, nasopharyngeal cancer and certain conditions linked to HIV infection, such as hairy leukoplakia and central nervous system lymphomas. Indeed, some authorities have worked out that as many as 200,000 cancers every year may be caused by, or linked in some way to, EBV infection.

Some researchers have suggested that infection with EBV increases the risk of autoimmune conditions, such as dermatomyositis, systemic lupus erythematosus, rheumatoid arthritis,

Sjögren's syndrome and multiple sclerosis. This is an ambitious litany, but, given the fact that EBV targets a key cell involved in human immunity, the B-lymphocyte, and given the very high prevalence of EBV infection – in the United States the virus is thought to infect roughly 50 per cent of five-year-old children and about 90 per cent of adults – perhaps the litany is not as unlikely as it first seems. But, as with CMV, this very high prevalence of infection with the virus should make us a trifle wary of ascribing causation unless we can demonstrate it beyond doubt, including the exact step-by-step pathogenesis.

In its structure, the EBV is typical of the human herpesviruses. Under the electron microscope, it has a capsid with the familiar icosahedral symmetry, wrapped in the baggy envelope of lipid and proteins. The virion is roughly 120–180 nanometres in diameter, and it is coded by a double-stranded helix of DNA which contains 85 genes. Its surface projections, made of viral-coded glycoproteins, are designed to discover and then interact with the cell membrane of specific human target cells.

Given the viral prevalence, it is inevitable that most people will sooner or later come into contact with EBV. In the majority of younger children infection appears to provoke no symptoms at all. Indeed, age of infection appears to be an important factor in predicting symptoms, if any. This perhaps tells us something about the virus–human interaction. But when acquired during adolescence, it provokes the typical symptoms and signs of glandular fever in some 35 per cent to 50 per cent of those infected. Like CMV, EBV takes the form of a long-term subclinical infection, with the virus being intermittently shed from the cells of the throat in individual carriers. Carrier status is common among adolescents, and the virus gets into the saliva, from which it is transmitted to new hosts through kissing. Perhaps it is not so surprising that glandular fever is mainly a disease of adolescents and young adults.

In the new host, invasion begins in the epithelial lining cells of the throat, when the viral envelope fuses with the cell membrane, after which the viral genome is transported to the cell nucleus. Here, the virus follows a similar pattern to the other herpesviruses, with the virus hijacking the nuclear machinery to produce daughter viruses. But the succeeding virus–host interactions are different; the released daughter viruses attract the attention of attacking B-lymphocytes, which would normally form antibodies to the virus. But in a second phase of infection, the virus now targets the B-lymphocyte, so that within the immune cells themselves the process of viral invasion and hijacking of the genomic machinery takes place once again. Within the B-lymphocyte there are now two possible outcomes: one is the so-called lytic pattern, which will inevitably rupture the cell with the release of daughter virions, ultimately giving rise to blood-borne spread of the virus. But there is a second, quite different potential outcome. In this cycle, the virus adopts a pattern of 'latency', in which it does not programme the lymphocyte to produce daughter virions. Instead, the viral genome assumes a circular form, known as an 'episome', which resides within the lymphocyte cell nucleus and is copied by the cellular DNA copying machinery during subsequent cell divisions. Again, we glimpse a potential for mutualistic behaviour in that such latency may contribute to future immunity against re-infection with the same virus.

The incubation period of glandular fever extends to a month or even longer. The illness usually presents with a fever, sore throat and enlargement of the lymph glands around the angles of the jaw. As the virus becomes blood-borne, other elements of the body's systemic defences take up the challenge and it is during this phase that liver function may be damaged and the spleen may become sufficiently enlarged for the examining doctor to detect this through feeling the tip on bimanual examination of the abdomen. Examination of the peripheral blood shows a characteristic rise in white blood cells, especially lymphocytes, which gives the disease its name, 'infectious mononucleosis'. Some patients develop a transient rash during this systemic phase. But serious complications are thankfully rare. These include the so-called Guillain-Barré syndrome, which presents with peripheral nerve damage and resultant paralysis, and, rarely, rupture of the enlarged spleen. Fortunately, most affected individuals go on to make a complete recovery within three to four weeks.

Viruses can be very strange and problematic. We do not, as yet, know why the EBV is relatively benign in the Western world, meanwhile it causes tumours in young Africans and epithelial cell cancers in southern China. Perhaps it reflects minor genetic variants in histocompatibility genotypes of different human populations, or perhaps different strains of the virus? In Burkitt's lymphoma, it seems that the strange latency described above is somehow involved. The wonder is that the tumours respond brilliantly to anti-tumour drugs, such as cyclophosphamide, which can bring about a complete cure. We should be thankful for the fact, that when it comes to glandular fever in the Western world, the latency in our B-lymphocyte cells appears to protect us from re-infection for the rest of our lives.

10

How Flu Viruses Reinvent Themselves

In the autumn of 1918 pandemic influenza erupted in Europe, America and parts of Asia, before sweeping through continents still reeling from the carnage of World War I. Despite the fact the epidemic was global, it came to be known as 'the Spanish Flu', reportedly because the media in Spain was not blanketed by wartime censors who otherwise deliberately minimised the reports of flu mortality in the United Kingdom, Germany, the United States and France, in order to maintain the morale among troops and citizens. We might baulk at such censorship today, but given the deadly nature of the pandemic and the fact that there were no vaccines to prevent its spread and no antiviral drugs to treat those suffering from the illness, it was perhaps understandable. We should add in the fact that the medical profession at that time knew very little about the causative virus of influenza, since DNA had yet to be discovered and the electron microscope had yet to be invented. This same lack of understanding led to a failure of the most basic measures to contain the flu's epidemic spread, judging from contemporary photographs, which show large numbers of sick soldiers filling open hospital wards, their beds crammed closely together, with no evident attempts at barrier nursing, and with patients and nursing staff devoid of simple face masks.

To suffer influenza at any time is decidedly unpleasant. To be caught up in a virulent pandemic when actively serving your nation in a war offensive must have been testing in the extreme. The Meuse-Argonne Offensive was a decisive battle in the Western Front, involving the largest frontline commitment in American military history, and engaging some 1.2 million troops. With more than 26,000 of the US troops losing their lives in the battle, it is also considered the single most deadly encounter in American military history. Unfortunately, the same offensive coincided with the arrival of the 1918 European flu pandemic. This struck the main US Army training camps, killing an estimated 45,000 soldiers: thus, in the words of authors Wever and van Bergen, 'making it questionable which battle should be regarded as "America's deadliest"'. Today historians regard the Spanish Flu as the most lethal influenza epidemic in history, infecting half a billion people worldwide and killing an estimated 20 to 50 million victims.

Most of us will have suffered a less deadly encounter with the influenza virus. I am not referring to the common-or-garden 'touch of the flu' employed as a jaded metaphor for a hangover. You are unlikely to forget those first two or three days of the real flu prodrome after contracting the virus, during which it is multiplying within your bloodstream. You might recall a feeling of imminent death? Perhaps your partner thought you were exaggerating – a situation rectified when said partner contracted the virus and got to experience the same dread. What you were feeling would have been much the same as those millions of unfortunate people experienced during the prodrome of the pandemic back in 1918. The difference is that you got better while for many of those suffering the Spanish Flu the dread proved to be prophetic. This difference in outcomes raises an important question: why did we survive and those millions of people die when infected with a similar virus?

Perhaps we should rephrase the question to put it into a more definable scientific language? What would cause the virus of 'ordinary' flu to behave with such terrible virulence? To address this question, we need to understand something of the history of influenza and the modus operandi of the causative virus.

Where, for example, did the term 'influenza' come from? It is no accident that the name of the condition resembles the English word 'influence'. In fact, both words come from the same Latin root, *influentia*, reflecting the fact that in medieval times a more superstitious population attributed epidemics to malign spells of astral or occult 'influence'. In these more enlightened times we can dismiss the magical implications and examine the true microbial cause of influenza – the influenza virus – in detailed anatomical, physiological and genetic detail. What is more, we can take an objective view on how flu viruses evolve in order to consummate their replicative symbiotic cycle with their natural evolutionary host – we humans.

The flu virus belongs to the family of the orthomyxoviruses. These comprise seven genera of RNA-based viruses, including four different genera known to cause influenza, which are conveniently labelled A, B, C and D. The first three of these cause flu in vertebrate animals, including birds, humans, pigs, dogs and seals, with influenza D limited to pigs and cattle. Humans are exclusively infected by influenza A and B. The individual virions are between 100 and 200 nanometres in diameter and are roughly spherical in shape. The surface of the virus comprises a lipid-based envelope which is covered by hundreds of projecting spikes. Changes in the proteins of these spikes are the root cause of new epidemics of influenza. The spikes are made of two different proteins, the haemagglutinin or H protein and the neuraminidase or N protein, which are the virus's means of attaching to the host target cell. The human immune system recognises the H and N proteins as alien 'antigens' and it manufactures antibodies to do away with them. The influenza A virus, for example, has numerous subtypes or 'strains' of both the H and N proteins. To clarify the terminology, a strain of virus labelled H2N28 carries the H2 and N28 subtypes of antigen proteins on its spikes. These subtypes change as a result of mutations in the viral genes, coding for them during replication of the virus. If the mutation gives rise to an increase in infectivity of the virus – more infectivity implies greater success in spread and replication – natural selection will 'positively select for' this evolutionary success. It is this process of evolutionary emergence of new subtypes or strains under the influence of natural selection that gives rise to new epidemics of flu.

It was, for example, the emergence of a novel H1N1 strain that caused the so-called 'Spanish Flu' of 1918; a novel H2N2 strain that caused the 'Asian Flu' of 1957; a novel H3N2 strain that caused the 'Hong Kong Flu' of 1968; and a novel H7N9 that gave rise to the 'Bird Flu' epidemic of 2013.

Some readers might confuse these flu epidemics with the respiratory infection known as 'severe acute respiratory syndrome', or 'SARS', which originated in China in 2002. In fact, SARS is not caused by an influenza virus but by a completely different virus known as the coronavirus, SARS-CoV. Prior to the SARS epidemic, coronaviruses, so-called because the virions are crowned with thorny projections on electron microscope views, were known to infect animals and birds, causing a severe cold-like illness. The SARS coronavirus put a new perspective on these viruses when, originating in Hong Kong, it caused a flu-like outbreak affecting some 8,098 victims in 37 different countries, including some 774 fatalities. Thankfully no further epidemics of SARS have been reported since 2004.

Like SARS, the flu virus is extremely contagious, through inhaling the aerosol projected into the ambient air from the coughing and sneezing of an infected person. Contagion is more likely in circumstances where people are brought into close proximity with one another, for example in airplanes, offices, public transport, schools, colleges and, of course, the family home. The virus is also infectious through saliva, nasal secretions, faeces and blood. These bodily fluids can contaminate surfaces which are then rendered infectious to others, usually through fingers coming into contact with contaminated surfaces, from which the virus is transmitted through contact with the mouth or the eyes. Frequent hand-washing in an infected household reduces the risk of spread, as does wearing a surgical mask when exposed to a coughing and sneezing victim.

The high degree of infectivity of flu viruses was illustrated by a mini outbreak on a commercial jet airliner in 1979. The plane, carrying 54 people, was delayed on the runway in Alaska for three hours, during which time the ventilation system was inoperative. A single passenger happened to be suffering from flu. Within a few days, 72 per cent of the other passengers came down with the virus. Once within the mucous membranes of your nose, throat and respiratory passages, the virus uses its surface spikes to attach to specific sialic acid sugars on the lining cells. After wrapping itself in the cell's own membrane the virus tricks the cell's own process of 'endocytosis' to ferry it into the cell's interior. Here, having shed its capsid, the virus does the usual virus thing, hijacking the cell's nuclear machinery to turn it into a factory for the manufacture of a new generation of viruses. With the release of the viral brood into the respiratory passages, the host cell dies. Meanwhile the burgeoning brood infects more of the lining cells, and so the process extends and deepens.

What the infected person experiences is an incubation period of two or three days, while this process of invasion and replication is amplifying, followed by rapidly escalating symptoms and signs of full-blown infection. There is often an abrupt onset of shivering, sickly feeling, headache and aching in the limbs and back. The body temperature rockets to around 39°C, which precipitates heavy sweating. This is the stage when we are prostrated and we feel that awful sense of doom. Ironically, once again these symptoms and signs are not the direct effects of the virus infecting our respiratory cells but are brought about by the ferocious reaction of our immune system in fighting off the infection.

The presence of the virus provokes white blood cells, known as macrophages and neutrophils, to invade the infected tissues. These cells sound the alarm by producing chemicals known as cytokines and chemokines that call in reinforcements, including T-lymphocytes, also known as 'soldier cells', to do battle with the viruses. Indeed, battle is the operative word, with these soldier cells now killing the body's own infected cells. The killing zone becomes highly inflamed, with an outpouring of mucus, which clogs the air passages and provokes sneezing and coughing. While the virus is normally confined to the air passages, the same chemical warnings enter the bloodstream and provoke the high fever, headache, the sickly feeling of fatigue and the uncomfortable muscle aches. A decline in T-cell function, such as occurs in old age or in immunocompromised patients, can make matters worse and prolong the illness, or albeit in a small minority, lead to a secondary bacterial pneumonia, which can be life-threatening.

The consolation is that the vast majority of sufferers of seasonal flu go on to make a full recovery. But we should not underestimate the potential for serious disease in that minority of sufferers, particularly so the elderly and the immunocompromised. In such circumstances there is an appreciable risk of hospitalisation, and a significant level of mortality. We should be thankful that in our modern world vaccines are available that can mitigate the disease, and, though not always as efficient as we would like them to be, can protect people from getting the infection in the first place.

Prevention through an efficient and rapidly dispensed vaccine is a highly desirable objective – but this depends on a deeper understanding as to how epidemics and pandemics of influenza arise in the first place. Changes in those surface spike proteins are the key to the emergence of the familiar winter epidemics of flu.

Our human immunity against the virus is based on antibodies raised against the two antigens in the flu spikes. The more the viral antigens have changed since our last encounter, the weaker the power of our antibodies to fight against the new strain. This is why even though we have previously suffered from flu, or have been vaccinated to deal with previous strains, this fails to protect us when a new strain appears in a subsequent winter. In the words of a leading expert, 'The genetic liability of the influenza viruses makes them capricious and formidable enemies to world public health.'

Pandemic flu is even more problematic. Fortunately, it is much rarer than the seasonal attacks, but when it does come along it is altogether more threatening. Once again it helps to understand the evolutionary mechanisms that lie behind such a perilous emergence. Pandemic flu does not arise from mutations in the H and

N spikes but through a far more powerful evolutionary mechanism. When two different flu viruses coincide in a single host, for example in a pig, they can swap whole chunks of their genomes to produce a new hybrid virus. Pandemic flu is exclusively confined to the influenza A genus. And since it gives rise to what, in effect, is a completely new virus, our human immunity is much less prepared to do battle with it than the seasonal viruses.

Will it ever be possible to eradicate the threat of pandemic flu through some special vaccination programme, like how smallpox was eradicated? I'm afraid that, although vaccines are likely to prove more and more preventive, and new antiviral drugs are also likely to improve our treatment of the disease, it seems unlikely that we will ever completely eradicate influenza. Smallpox was eradicated because humans were the sole reservoir of the virus – but humans are not the sole natural reservoir of influenza viruses. The natural reservoirs of influenza are the aquatic birds of the world. Wild ducks and other water fowl already harbour some 14 different H antigens. I'm afraid that this means that the potential for new pandemic strains of influenza already exists in this natural gene bank. All of these different influenza viruses replicate in the digestive tracts of wild fowl, which excrete them into the aquatic ecologies they inhabit. For example, when scientists drew samples from the great lakes of Canada during the winter, they found extensive contamination with different populations of influenza virus. And yes, we are looking at a situation we have seen again and again with other viruses in relation to their natural hosts. When scientists examined the birds that are the influenza virus's natural hosts, they found that the influenza viruses caused no apparent disease.

Some years ago I called in on Nancy Cox, then Director of the Influenza Division at the Centers for Disease Control, in Atlanta, to talk about the future risk of pandemic flu. In her words, 'In this sort of instance, with a population lacking immunity and a virulent pandemic strain on the move, we have a very dramatic situation developing.'

On the wall of her office, Dr Cox had maps of the world, decorated with spreading contour lines and various coloured pins. Like other influenza experts around the globe, she was attempting to predict when and where a new pandemic flu might emerge. She believed that past behaviour might give her an idea of what to expect. Experts who have to deal with pandemic strains of influenza spend much of their time examining strain behaviours and tracking virus evolution in this way.

On the lower part of her wall was a map of China, with six different sites ringed. All these were being monitored by observers hoping to pick up the new strains as soon as they arrive. But China is not the only possible site of emergence and there are observers monitoring the situation in other destinations worldwide. In 2017, the H7N9 Bird Flu returned to have its deadliest year since it first arrived in 2013, causing serious illness in 714 people in China, with a reported mortality of more than one in three. If and when a new pandemic strain of influenza virus emerges, a dramatic race will begin. The aim will be to incorporate the appropriate antigens of the new virus as quickly as possible into a preventive vaccine. With the pandemic strain the world would have no more than months to prepare and distribute sufficient new vaccine in advance of a virus that will circumambulate the globe at the speed of a passenger jet. Speed and accuracy of prediction will then be the key in what will become a global matter of life and death.

11

A Lesson from a Machiavellian Virus

A decade ago a previously healthy 73-year-old Canadian man presented to his local hospital with a pain in his shoulder that was followed by fever, difficulty in swallowing, muscle spasms and generalised weakness. His neurological status deteriorated, and he showed increasing irritability, lethargy and the rather unusual complication of excessive salivation, to the extent that he was leaking saliva out of his open mouth. Two days later his limbs and body began to spasm and jerk and he lost his mental faculties before entering a state that neurologists call 'decorticate posturing'. Decorticate, in this sense, signifies the loss of function of the cerebral cortex, the site of higher function in the brain.

Normal resuscitation measures were conducted, with intubation and mechanical ventilation of his lungs. Fluids were injected into his veins, and drug therapy including antibiotics and steroids was administered in desperate attempts at resuscitation. A CAT scan of his brain was surprisingly normal. But his treating physicians had developed a growing suspicion as to what was really going wrong inside their patient. They asked the man's relatives if he had been exposed to an animal bite. The family confirmed he had indeed been bitten by a bat on his left shoulder six months previously but had not considered it necessary to see a doctor for treatment. The medical team took a biopsy of the skin of his neck and they took samples of his saliva and further blood samples, all now focusing on a single diagnosis. The laboratory results confirmed their fears. The patient was in a terminal stage of rabies. The doctors initiated treatment with what is known as the 'Milwaukee Protocol' but it was far too late to save this unfortunate patient and, although he survived in a comatose state for two more months, he was confirmed brain dead and passed away when the medical support was withdrawn.

The autopsy revealed purulent viral meningitis and a microscopic examination of the cerebral cortex confirmed that the rabies virus had destroyed all of his brain cells in the parts responsible for higher mental function.

This poor man's case was all the more tragic for the fact that the outcome might have been so very different had he presented for help soon after he had been bitten by the bat. It might appear astonishing that the single bite of a bat had killed him in such a dreadful fashion, but in fact this man's nightmarish decline would have been altogether familiar to the authors of the Babylonian Eshnunna Code some four thousand three hundred years ago. Those ancient authorities decreed that if a mad dog were not brought under control by its owner, and if it should bite a man and cause his death, the owner of the dog would be fined 40 shekels of silver.

For thousands of years untreated rabies continued to be fatal, although the Roman sage, Aulus Cornelius Celsus, recommended cautery of rabid animal bites with a hot iron, a harsh treatment yet one that might, if applied immediately after the animal bite, have proved curative. Few appeared aware of this drastic remedy and rabies remained a fatal scourge until the pioneering French microbiologist, Louis Pasteur, introduced the first rabies virus vaccine in 1884. Even today, the Pasteur vaccine is our best hope of preventing this terrible illness in those who have been bitten by a rabid animal. But these days we have additional measures that might help save the life of an infected individual – provided we catch the disease at an early stage.

The rabies virus is one of the strangest and most virulent of the known viruses, adopting strategies of seemingly Machiavellian cunning to enhance its survival and propagation. It is a member of the genus of lyssaviruses, a name derived from the Greek *lyssa*, which translates as 'frenzy', and which describes the nature of the madness induced by the virus in an infected animal, or sometimes in a human. Why would a virus evolve such a terrible strategy that would result not only in the annihilation of its victim, but also of the populations of virus within the same victim? The answer is likely to be complex, but at least in part it may reflect the fact that the victims are not the virus's natural hosts.

Rabies and all lyssaviruses are members of the family of rhab-doviruses, which infect an extraordinary range of hosts, including reptiles, fish, crustaceans, mammals and even some plants. In the opinion of Hervé Bourhy, an expert on the subject at the Pasteur Institute, the rabies virus is the symbiotic partner of bats, in which it causes no signs of disease. But the same virus is capable of 'infecting' a wide variety of mammals, including foxes, coyotes, jackals, rodents and, of course, dogs, all of which can be considered expendable from the evolutionary perspective. Thus the seemingly suicidal mortality of the viral infection in non-host species of animals, including humans, poses no threat to viral survival, since the virus will continue its existence in the bat host. That being so, it would be hard to conceive a more sinister modus operandi than that of the rabies virus in its less favoured prey, which is programmed to infect the centres in the prey animal's brain that induce uncontrollable rage, while also replicating in the animal's salivary glands to best spread the contagion through the rage-induced frenzy of biting any living creature that comes within range. It is hard to detect long-term evolutionary purpose in this, other than to eliminate ecological rivals or potential threats to the natural host. Perhaps we should ask ourselves a different question: how does the virus spread among bats?

Bats constitute the second commonest mammalian order, with approximately 1,200 constituent species. It seems unlikely that rabies viruses will be symbiotic with all of them. We just don't know enough of the rabies virus's symbiotic partners among the bats to answer such a question. But what if the evolution of lethality associated with biting evolved not in relation to dogs or humans, or indeed any of the above-mentioned mammals, but as a strategy within the competitive world of the different species of bats? Competition for living spaces and resources might then make the Machiavellian behaviour altogether comprehensible.

The family name, rhabdoviruses, is derived from the Greek *rhabdos*, meaning rod. Rabies virus is indeed rod-shaped, with one end flat and the other rounded, which actually gives it the shape of a submicroscopic bullet. The virion is 170 nanometres long, ensheathed in a lipid envelope that protects the viral capsid, which in turn protects the genome. Where the viral capsids we have encountered to date have exhibited the crystal-like icosahedral symmetry, the capsid of the rabies virus, as with all rhabdoviruses, twirls around the genome in a helical symmetry. Other than bats, rabies is probably capable of infecting all warm-blooded animals but there is a hierarchy of victim susceptibility, with foxes, coyotes, jackals and wolves being the most susceptible. Surprisingly, dogs are only moderately susceptible, but, as a consequence of their close relationships with people, they are the commonest vector for transmission to humans worldwide. This risk can be mitigated through well-organised veterinary programmes of canine vaccination: but this would not prevent the risk from bat bites, or some of the other potential secondary sources, which include cats, and in the North American continent animals such as raccoons and skunks.

When the rabies virus enters a susceptible prey, whether through the bite of its non-rabid bat host or through the bite of a rabid animal, the bite carries the virus through the cutaneous barrier, allowing access to the deeper tissues. It seems that even the simple licking of an abraded skin with infected saliva may transmit the infection, as can infected saliva coming into direct contact with the eyes, mouth or nose. After transmission, the incubation period varies from ten days to a year or more. After the bite the virus first multiplies in skin and muscle cells before finding its way to the peripheral nerves, through which it ultimately travels to the brain. Here the virus discovers its ultimate target cells: the nerve cells that populate the upper echelons of the brain, which, in humans, determine higher mental functions.

Pain and tingling at the infected site, accompanied by jerky local movements, are suggestive of incipient disease. But only when the virus enters the brain does the patient show the classical symptoms and signs of rabies. These follow one of two patterns. In a minority of sufferers there is a gradual ascending paralysis of the body, ultimately leading to death. But in some 80 per cent of patients, rather as happened in the Canadian gentleman, the sufferer enters a stage of excitement, with anxious and apprehensive facial expression, a fast pulse and rapid breathing. In one reported case in the UK, such was the psychological disturbance that the patient's condition was initially misdiagnosed as schizophrenia. Spasms and paralysis of the facial muscles, and similar tic-like symptoms elsewhere, are typical of this stage, as is the onset of the pathognomonic hydrophobia. The pitiable individual feels a desperate need to drink, but when he or she attempts to do so, even the sight of water provokes violent spasms of the throat and respiratory muscles, accompanied by feelings of overwhelming terror. After a week or so he or she suffers widespread paralysis, coma and cardiovascular collapse.

Thankfully rabies is now potentially preventable and even treatable after infection. Vaccination of dogs, and people considered to be at increased risk of the disease, has been a considerable success, for example, in North America, Australia, Japan and much of Western Europe. Even in people who have suffered a bat bite, or the bite of a suspect animal, treatment with rabies immunoglobulin can avert the disease – but this needs to be administered within ten days of infection. How sad, therefore, that even today rabies remains a widespread illness throughout the world, killing some 17,400 people in the year 2015. The majority of these deaths were in Africa and Asia, with some 40 per cent of them involving children under the age of 15 years.

This dangerous virus is unlikely to go away while its natural host, the bat, survives. We have witnessed much the same phenomenon in relation to other epidemic and endemic viruses, for example, with influenza in relation to water fowl, and the hantavirus in relation to the deer mouse. This prompts a relevant and important question: how do such virus–host symbiotic relationships arise in nature? To answer this we might look for inspiration in an instructive bio-warfare experiment involving a man-made plague of rabbits.

*

The wild European rabbit was first introduced into Australia in 1859 as a source of food for European settlers. Lacking natural predators, the rabbit population underwent an explosive expansion, leading to widespread destruction of agricultural grassland. Between March and November 1950, in a deliberate act of biological warfare, the Australian authorities infected feral rabbits with a virus with the intention of decimating their numbers throughout the territory. The virus they chose for these purposes was a pox virus of the Brazilian wood rabbit. This virus is responsible for a persistent infection in the Brazilian rabbits, spreading through biting insects. It is important to grasp that in this case persistent means that the virus never abandons its host, whether at individual or species level, after its first arrival. Such a situation encourages the development of what virologists call 'co-evolution' of virus and host, an alternative term for an ongoing symbiotic interaction.

While the rabbit pox virus causes little in the way of illness in the Brazilian rabbit host, some strains were known to be exceedingly lethal for the European rabbit, where they caused an illness known as 'myxomatosis'. Between March and November 1950, the Australian biologists inoculated test groups of wild European rabbits at five field sites in the Murray Valley of southeast Australia with a highly lethal strain of the virus. It was not planned as an evolutionary experiment, but in retrospect it is seen as a test model with important evolutionary implications. In effect, the circumstances replicated the situation that might have taken place if a species of rabbit already co-evolving with the virus in the wild might have come into contact with a species that had never encountered the virus. In such circumstance, the virus is apt to jump species to give rise to the pattern of an 'emerging virus infection'.

Emerging viruses are an important source of epidemic and pandemic plagues. Some examples from recent history include HIV-1, Ebola, the *Sin Nombre* hantavirus, Lassa fever, bird flu, SARS and, most recently, the Zika virus. Biologists had previously attempted similar experiments involving myxomatosis virus in Australia and Europe, but every one of them had failed in its intention of controlling the burgeoning rabbit populations. As this new experiment got underway, the Australian scientists were not altogether surprised when, once again, little appeared to happen. The virus appeared to spread poorly throughout the nine months or so of the inoculations. But those nine months were relatively dry and the myxomatosis virus spreads through biting insects. Suddenly, during December, with the proliferation of mosquitoes following a wet spring, an epidemic exploded. The myxomatosis virus preferentially targets the cells of the rabbit immune system – a pattern that is now becoming somewhat familiar from the examples of many different viruses.

The initial target cells of the virus were the major histocompatibility type II cells in the rabbits' skin, from which it spread to the neighbouring lymph glands and, further, through blood spread to the spleen, where the virus targeted the white blood cells known as lymphocytes. Here the viruses replicated and spread until they were now teeming to levels of a hundred million in a single gram of infected lymph glands or spleen. From there, further bloodstream spread resulted in an almost universally lethal illness, with affected rabbits showing pathologically swollen heads, eyelids and ears, raised sores over the skin of the body, ears and legs, pronounced ano-genital swelling, marked inflammation of the membranes of the eyes and bloody nasal discharges. The virus that caused little in the way of illness in its symbiotic partner, the Brazilian rabbit, proved apocalyptic to the Australian rabbits.

Within three months of the onset of the epidemic, 99.8 per cent of the rabbits of southeast Australia, a land area the size of Western Europe, were exterminated by myxomatosis. The observing scientists were witnessing up close the potential of virus–host evolutionary interaction in the red-in-tooth-and-claw of Darwinian natural selection. But this was not quite the way it would have operated in nature without the deliberate hand of humans. The natural clash would have involved the myxomatosis virus arriving into the rival rabbit's ecology in the company of its natural host, the Brazilian wood rabbit, when we might anticipate this level of viral aggression to have 'culled' the Australian rival, thus making way for the Brazilian rabbit to dominate the ecology. This is a classic example of evolutionary mechanism I have termed 'aggressive symbiosis'. But in this human-induced scenario there was no evolutionary rival to take advantage of the cull. This radically changed the evolutionary dynamic. While the Australian authorities might have hoped for the total extermination of the feral rabbit population, this did not materialise. Instead an alternative dynamic, following a different pattern of aggressive symbiosis, duly emerged.

The resistance to infection, like the determination of 'self', is determined by the chromosomal region known as the 'major histocompatibility complex'. In rabbits, as in any mammalian species, there are genetic variants that have evolved in various populations of rabbits that are important in determining the response to infection. These are known as 'genotypes'. We can assume that, within the overall rabbit population, the myxoma virus would be more lethal for some genotypes than for others. Thus the epidemic, caused by a persistent plague virus, would inevitably cull the genotypes that were most susceptible to lethality within the overall rabbit population. The very high mortality suggests that the majority of rabbits had a genotype that was highly susceptible. But this same culling effect would have a lesser effect on a minority genotype of the rabbits that was less susceptible to fatality.

Histocompatibility genotypes are hereditary. Generation by generation – and here again we see the importance of the virus being persistent – selection for resistance to lethality would have been ongoing. In time, and given that rabbits reproduce quickly, a new rabbit species genotype emerged that could survive the persisting presence of the virus. The myxomatosis virus had now selected a new host.

Culling was followed by symbiotic co-evolution between the virus and its new host so that, within just seven years, the lethality of the virus in its new partner was reduced to 25 per cent. This symbiosis between rabbit and virus has continued to the present day, when the now rapidly re-expanding rabbit population is almost completely resistant to the former lethality of its aggressively symbiotic partner. The virus's evolutionary needs, of survival and reproduction, are satisfied by this relationship with its new host and partner. But what, one might wonder, is the gain from the relationship for the rabbit? We need only consider what would happen if a rival species of rabbit, virgin to the virus, were to enter the ecology of the co-evolving partnership . . .

Not only have we witnessed how an aggressively symbiotic virus would protect its host from an ecological rival, we have witnessed how the same aggression would enable a virus to establish a stable symbiotic relationship with a new host. The manner of such a developing symbiotic relationship might appear ruthless, but this implies a moral judgement. Viruses are exclusively driven by primal evolutionary forces: survival and replication. They are quintessentially amoral.

12

The Mystery of Ebola

On 27 June 1976 a man living in the town of Nzara, in the far south of Sudan close to the border with the Democratic Republic of the Congo, fell sick. Nzara had only recently been hacked out of the tropical rain-forest hinterland. Colobos monkeys still lived in the trees and colonies of baboons still squabbled for territory in the high grass. Nzara had a population of 20,000 people, mainly of the Azande tribe, with family taboos, polygamy and ancestor worship. There were a few brick buildings, roofed in corrugated iron, but most residents, including Yusia and his two wives, lived in communities of mud-walled and thatched 'tukels'. On 27 June Yusia fell sick. A pain began over his forehead and spread until it involved the whole of his head. It was nauseatingly severe. Next he developed an extremely sore throat, which he described to his brother Yasona as a ball of fire. Slim and fit, Yusia had never been seriously ill in his life, but now his tongue was as dry as rope, crops of painful ulcers stung his cheeks and it was agony to swallow saliva. Soon after this he suffered severe muscle pains in his chest, his neck and the small of his back, from where the pain radiated down both his legs. His face became sunken and drained of expression. He just lay writhing and groaning on his bed. Yasona stayed with him in his mud-and-thatch home, nursing him through an illness that was progressing with a mind-numbing intensity and rapidity.

No measure seemed to alleviate Yusia's suffering. By 30 June his brother was so alarmed he arranged for him to be taken to the local hospital.

The hospital in Nzara was a shack with a few iron-frame beds. There was a dispensary run by a nurse and a locally based doctor who spent much of his time hunting monkeys. By now Yusia was suffering cramping abdominal pain, diarrhoea, vomiting and pros-tration. Two days after his admission, he started to bleed profusely from his nose and mouth and the diarrhoea became heavily blood-stained. The flesh shrank about his bones until his face resembled a skull, with eyes that were sunken and staring. Death came as a merciful release on 6 July.

Mortality from a febrile illness is commonplace in the tropics. Malaria, typhoid, tuberculosis and sleeping sickness have a firm clutch on the people and any one of these might cause a similar presentation and fatal outcome. But this illness seemed different from the usual run of ailments, and soon other people also began to fall sick with those same unbearable symptoms. Yusia would prove to be the index case of an emerging fever that was spreading rapidly in the local population. In time it became evident that the fever was spreading through close physical contact. In Yusia's community, the sick are traditionally tended by their wives and close relatives and the bodies of the dead are manually washed and cleansed by those same close family members prior to burial. The bathing ceremony is accompanied by an open display of grief, with weeping and fondling of the body and kissing of the face of the dead. Such intimate contact now led to an explosive local transmission. By August the contagion had spread to Maridi, a town with a large and well-equipped hospital, 80 miles east of Nzara. Here the fever spread among the old-fashioned Nightingale wards, with their columns of beds all in close proximity to one another, killing patients and staff with equal impunity, before invading the neighbouring Democratic Republic of the Congo, then known as Zaire, where in late August it struck the hospital in a town called Yambuku. Some 825 kilometres southwest of Maridi, Yambuku was located in the Bumba Zone, which straddles the Equator. In 1976 this zone encompassed an extensive population of some 275,000 people. Here, in a relatively well-organised Catholic mission hospital, the fever again exploded.

The nature of the fever was baffling. There had been talk of yellow fever, but few if any of the patients were jaundiced. While the illness had many of the features that might commonly accompany any of the familiar fevers, there were strange manifestations, symptoms and signs that were different in scale or simply bizarre. The patients were visibly terrified by what was happening to them. They were bleeding from every orifice. Some became confused and agitated, as if their brains were addled, causing them to strip off their clothes and climb out of bed. As the sickness deepened, their faces became strangely expressionless, their eyes sunken and glazed, like a mask of impending death. By now the illness was killing the nurses and doctors, too. It appeared to be unstoppable, spreading to involve everybody who tended patients. The native staff were so terrified they abandoned the stricken hospital. Nothing the dedicated nuns could do seemed to give the slightest relief. Nobody dared to wash or bury the dead any more. Even the very prisoners from the gaols refused to move the bodies, in spite of the promise of liberty. And the reports filtering through from the surrounding villages were equally alarming.

It was inevitable that panicking local doctors should complain to the Minister of Health in Kinshasa. Anguished calls for help went out to Brussels – the Democratic Republic of the Congo had formerly been a colony of Belgium, when it was known as the Belgian Congo. At the same time reports of the horrifying local situation caught the attention of the World Health Organization, in Geneva. There was now a growing fear that the epidemic might herald a newly emerging virus. Epidemiological and virological experts were commissioned to fly into Maridi from Belgium and into southern Sudan from Britain, tasked with discovering whatever was causing the terrifying epidemic.

In 1976 emerging viruses could be difficult to diagnose. There were a number of ways in which investigators might proceed. You could begin by looking for specific antibodies to the virus in the blood of infected patients. Such testing was rudimentary but it demanded a reliable viral antigen against which to test a patient's serum. The problem with a newly emerging virus is that you have no reliable antigen and therefore the standard serological tests prove useless. In such plague circumstances, there are two other ways of hunting for a novel virus. You could inoculate experimental animals with infected blood, or tissues, then look for pathological effects in the animals: or you could attempt to breed the virus in various types of cell cultures. When the experienced Brussels-based virologist Professor Stefan Pattyn began to investigate the blood and liver samples from a dead nun at the mission hospital at Yambuku, he adopted a first-line hypothesis. He thought it likely that they were dealing with Lassa fever, a haemorrhagic fever that had first broken out some years earlier in another African locality. So he instituted the appropriate investigation for Lassa, which involved inoculation of infected serum into mice.

There was an alternative and rather more worrying possibility: the agent might be a haemorrhagic fever virus but not the Lassa virus. If this were the case, Pattyn queried the possibility of an arbovirus. Arboviruses are viruses transmitted by biting insects – the term derives from 'arthropod-borne virus'. To cover this possibility he inoculated some more of the nun's serum into suckling mice, while homogenising some of the nun's liver and adding it to cell cultures in flasks. Next he fixed a sample of the liver in formalin and sent it to his colleague, Dr Gigase, a pathologist, who would subsequently examine the effects of the virus within the organs and body tissues of the inoculated mice. Within 24 hours, the pathologist phoned Pattyn to say that the liver showed a pattern consistent with hepatitis. Pattyn's heart jumped when the pathologist told him that he could see 'inclusion bodies' in the liver cells: inclusion bodies suggested a virus. But many different viruses attacked the liver. It was hardly diagnostic.

On 5 October, Pattyn called up Dr Paul Brès, Chief of Viral Diseases at the World Health Organization, in Geneva. Pattyn was surprised to discover that Brès already knew about the mystery African epidemic. Indeed, Brès now informed Pattyn that the WHO was extremely worried about it. However, the WHO Chief thought that Pattyn's lab was no better equipped to deal with a dangerous virus than an ordinary hospital diagnostic laboratory. The developing situation was too dangerous for routine lab precautions and he instructed Pattyn to send on all of his specimens to Porton Down in England, a biosafety-level-4 laboratory where they had the appropriate facilities to deal with exceedingly contagious and highly lethal viruses.

Over the next few days, Pattyn carefully bundled up the serum, the liver biopsies, some brains he had removed from the infected mice and his cultures, and he forwarded all of these on to Porton Down. By this stage Porton Down had also received specimens from the Sudan epidemic. Here an experienced virologist, Ernie

Bowen, organised a wide range of serological screens, cell cultures and animal inoculations on the specimens from both African epidemics, screening for a variety of potentially lethal haemorrhagic fever viruses, including yellow fever, Congo-Crimean fever, Rift Valley fever and Lassa fever. Bowen had a close working relationship with David Simpson, one of the most experienced field virologists in Britain, who was currently based at the London School of Hygiene and Tropical Medicine. Bowen picked up the phone and called Simpson to talk about the possibility that, as Pattyn had also wondered, they might be dealing with the Marburg virus. The two British virologists had worked together in the first clinical diagnosis of the Marburg virus some nine years earlier when the most lethal haemorrhagic fever to date had exploded into medical and public consciousness.

Back in the 1960s monkeys were treated as lab animals suitable for medical and biological experiments. The Marburg virus had been discovered in 1967 when a mysterious contagion had broken out in monkey-handlers at a factory in Marburg, Germany. Soon more human cases turned up in Frankfurt, and also in Belgrade, Yugoslavia. The discovery of the causative virus had sent a shockwave through the contemporary world of virology. Nothing remotely like this virus had ever been seen before. As the illness progressed, in both monkeys and human handlers, there was a vivid rash which tended to coalesce into a livid reddening of the skin over the face, the trunk and extremities. After the sixteenth day, the skin peeled off, the victim's hair fell out and they lost their nails. Seven of the infected people suffered severe haemorrhages from the nose, gums and from sites in the skin where blood had been drawn or where intravenous drips had been set up. They also vomited blood and passed blood from their bowels. In the most severe cases, particularly those who ultimately died from the infection, they entered a phase of confusion and coma, which signalled viral infection of the brain and meninges. These symptoms were remarkably similar to what was being reported from the patients in Maridi and Yambuku.

At Porton Down, after extensive experimental testing, David Simpson, then working as senior assistant to the chief virologist C. E. Gordon Smith, was the first scientist in the world to actually see the viruses that caused the Marburg epidemic. Simpson would later tell me how he could scarcely believe his eyes at what he was looking at through the enormous amplification of the electron microscope. In some views he saw what looked like horribly writhing snakes or worms. In other views they formed rings, like doughnuts, or alphabetical shapes, like question marks or commas. The bizarre viruses were about 80 nanometres thick but as much as 14,000 nanometres long in their serpentine coils. He turned to the technician working the electron microscope and demanded to know:

'What the hell is this?'

The British virologist had discovered a hitherto unknown family of viruses, now known as the filoviruses – from the Latin *filum*, which means a thread. They are amongst the most dangerous pathogens on Earth.

Nine years later, with the arrival of the specimens from Maridi and Yambuku, Ernie Bowen was now repeating the very tests they had employed during the Marburg epidemic. At the same time, virologists at the biosafety-level-4 lab at the Centers for Disease Control in Atlanta were equally interested in what was going on in Sudan and the Democratic Republic of the Congo. They included Fred Murphy, who was Chief of the Branch of Viral Pathology, and Karl Johnson, who was heading the Special Pathogens Branch. But there was little that CDC could do if they were not officially requested to intervene. When rumour suggested that Porton Down was involved, Johnson put in a call to Bowen. At the time Bowen was under an embargo of confidentiality from the WHO, but he admitted to Johnson that he suspected Marburg but could not as yet definitively prove it. Johnson offered him the assistance of an immunofluorescence test developed in America that would greatly speed up the testing. This made sense to Bowen, who sent Johnson serum derived from the dead missionary nun, together with some material from her liver biopsies, all of which arrived in Atlanta on 10 October.

The samples were tested by Johnson's wife, Patricia Webb, then working as his assistant in Special Pathogens. Within two or three days, she found that the serum from the deceased nursing sister provoked cytopathic changes in cultured cells, suggesting the presence of a virus. She passed on a sample from the liquid supernatant in the culture flasks to Fred Murphy, so he could examine it under the electron microscope. In Fred Murphy's own words: 'I stuck it into the microscope. Virtually immediately I saw long curled filaments, absolutely unique among all the viruses. It looked exactly like Marburg. Nothing else looks like that. The hairs stood up on my neck.'

They had discovered a second and equally deadly filovirus, which was named after a headstream of the Congo River: they called it the Ebola virus. The overall mortality of the 1976 Ebola outbreak in Sudan was 53 per cent and in the Democratic Republic of the Congo a horrifying 89 per cent. Subsequent to this there were a number of recurrences of Ebola infections in both countries but no major outbreaks until 1994, after which outbreaks of Ebola in various African countries have repeatedly made their way into the global news, including the most calamitous to date, in West Africa in 2014. Study of the responsible strains of virus has extended the label 'Ebola' to a genus within the family of filoviruses. The *Ebolavirus* genus now includes a diversity of species that pose significant threats to public health. These include Ebola Democratic Republic of the Congo, Ebola Sudan, Ebola Reston (this is actually an American story, but the virus appears to have originated in Asia), Ebola Taï Forest and Ebola Bundibugyo. The 2014 West African epidemic infected some 28,000 people and took the lives of more than 11,000.

Health authorities are considering global surveillance strategies that might predict the danger of new devastating outbreaks of Ebola and Marburg viruses and, hopefully, prevent future illness and mortality. But some important questions remain. One such question that has taxed virologists since the viruses first declared their deadly presence in Marburg, Germany, and Maridi and Yambuku, in Africa, is where do the viruses come from? To be specific, what is the reservoir of the filoviruses in nature?

Today we have an answer to that question, and one that helps link Ebola and Marburg to what we have been observing with other epidemic viruses. Fruit bats appear to be the natural host of the filoviruses. It is interesting to discover, yet again, that in their partnership with bats these terrifying viruses appear to co-exist without causing any signs of disease. This means that the viruses do not rely on their infection of humans, or other animals, to survive. They are guaranteed survival while ever their symbiotic natural hosts, the fruit bats of two continents, themselves survive. How the viruses come to be transmitted from bats to trigger human, or other mammalian epidemics, is still being investigated. One wonders if the lethality might be an advantage to their hosts as part of the competition for survival and ecology between bat species in nature. The patterns of the human outbreaks suggest that one or otherwise small numbers of human become accidentally infected when humans enter the ecology of the virus–host association, which, given the nature of the ecologies involved, would suggest that this happens mainly in rain-forest or wooded areas. Meanwhile, unlike viruses such as the *Sin Nombre* hantavirus, or rabies, the fact that the filoviruses are readily transmissible from human to human is worrying.

The Ebola story has additional worrisome implications. It would appear to have significant commonalities with other emerging viruses, such as the hantavirus, Lassa virus, HIV and SARS, in that these dangerous emerging infections appear to result from the rapidly expanding human population increasingly coming into contact with virus–host associations in nature. These natural reservoirs are called zoonoses. A study of such zoonoses suggests that patterns of viral diversity in wildlife might help us to anticipate future epidemic threats and thus enable us to take measures to prevent them. Bats harbour both the filoviruses and rabies. They also host Hendra and Nipah viruses, Asian-based biosafety-level-4 viruses that cause serious and potentially lethal infections of horses, pigs and humans. Is there a reason why we should be particularly wary of viruses coming from bats?

One possible explanation for the seemingly high association of bats with lethal viruses may result from the wide diversity of bat species when compared to other mammals. But a group of scientists arrived at a contradictory conclusion. After gathering all the information they could find on all the viruses known to infect mammals, they tallied some 586 different viruses infecting some 754 mammalian species. Using this data they devised a system that would allow them to calculate the 'viral richness' of each separate mammalian species. Next they assessed the latent pathogens that might cross species to infect humans. They concluded that, given the high diversity of bat species, coupled with the fact that each individual bat species harboured an average of 17 viruses, which was far higher than any other mammalian order, bats were more likely to harbour viruses that could cross to humans than any other wild mammal group.

Should we be paranoid about bats? The answer is: no! The majority of humans have little interaction with bats in nature – and though they are host to many viruses, they rarely seem to transmit their viruses to us. But more work needs to be done in terms of looking more closely at the precise nature of the symbiotic interactions between bats and their viruses. This should then inform scientists involved in future surveillance of viruses arising from zoonotic reservoirs how best to look out for viruses coming from bats, and indeed other, natural reservoirs.

13

The Mercurial Nature of the Zika Virus

Athletes travelling to the 2016 Olympics in Rio de Janeiro found themselves facing an unexpected threat. Earlier that same year the World Health Organization had declared a global public health emergency in relation to a virus, known as the Zika virus, which was producing severe developmental defects in babies born to mothers infected with the virus. It is likely that the athletes, like most people, had never heard of Zika, but this was about to change as the virus began to capture newspaper headlines throughout the world. What then was this strangely-named virus? Where had it come from? And why, all of a sudden, was it causing such a media panic?

In fact the Zika virus had been discovered as long ago as 1948, when it showed up as an unknown entity in the mashed-up bodies of mosquitoes of the *Aedes* genus that had been collected by virologists conducting a routine sweep in the Zika forest of Uganda. When serum taken from some of the local human population was tested for antibodies to the virus, Zika was found to be a hitherto unsuspected infection in humans. They had discovered a new infectious arbovirus.

Arboviruses are not a specific family of viruses, rather they include a medley of viruses from a number of different families. What they have in common is, of course, the capacity for transmission through the intermediary of biting insects. There is also a good deal of overlap in the clinical syndromes they cause. Zika was registered as a new member of the 'flaviviruses', a group with malevolent epidemic potential. It includes such notorious examples as yellow fever, West Nile virus and dengue fever. Further research showed that, although the virus was isolated from a number of different *Aedes* mosquitoes, the main vector was the female *Aedes aegypti*, which is active during the daylight hours and is obliged to gorge on fresh blood in order to lay her eggs. Compared with some of its infamous flavivirus relatives, Zika seemed relatively benign. When first studied in Uganda, humans appeared to be unintended victims of a virus whose natural host appeared to be forest monkeys and apes; and even when they became infected, humans suffered no more than a mild fever, sore eyes, some joint pain, headaches and a spotty rash. But there was an additional and worrisome surprise in store. When epidemiologists investigated the manner of spread of the virus in the local population, they discovered that, having gained access to the local human population through mosquito bites, the virus had evolved the capacity for spreading from human to human through sexual intercourse, childbirth and blood transfusion. The surprises did not stop there.

Soon this endemic virus of the Zika forest began to spread out of Africa and into Asia, invading a thin but spreading equatorial belt. It provoked little attention for six decades before manifesting in epidemic form, in 2007, on the tiny island of Yap in the Federated States of Micronesia. Here some 5,000 people were infected, amounting to around 70 per cent of the island's population. The illness never became life-threatening, spreading through the population over a few months after which it appeared to peter out. But the virus had not actually gone away. Zika broke out again in 2013, this time in French Polynesia, where it infected an estimated 30,000 of the population. The majority of infections were asymptomatic and even those with clinical illness suffered relatively mild symptoms. From here it spread to seven other island nations, where it infected small numbers and caused no deaths. But once again the virus changed its behaviour as, for the first time, albeit only in a minority of the infected, it caused serious neurological complications.

These complications included 42 cases of the Guillain-Barré syndrome, that same paralysis of the peripheral nerves that we saw as an occasional complication of the Epstein-Barr herpesvirus. These patients required lengthy hospital admission, and 12 patients needed ventilator assistance because of paralysis of the muscles involved in breathing. Some 43 per cent of these victims were left permanently handicapped with various long-term effects of paralysis. Zika virus infection could no longer be treated as benign. Meanwhile the virus was expanding its geographic territory eastwards across the Pacific Ocean, invading New Caledonia, Easter Island, Cook Island and Indonesia, and with first cases now registered in Australia and New Zealand in 2012.

By 2015 Zika was epidemic in the Americas, including Brazil. Early the following year it invaded North America and by early 2016, the World Health Organization warned that it was likely to spread throughout most of the remaining American territory. Medical authorities were alarmed by yet another change in the virus's behaviour. In addition to the growing variety of neurological complications, Zika was crossing the placenta in pregnant mothers to damage the development of the brain in the foetus. Distressing pictures of babies born with small heads, or microcephaly, began to appear on the front pages of newspapers. That same year sexual transmission of Zika virus infections was documented in the United States. This prompted the CDC to issue travel guidance for Americans travelling to affected countries. The guidelines warned that there was no effective antiviral drug to cure Zika virus infection. Instead they gave practical advice on reducing the risk of infection, together with a specific warning to pregnant women of the dangers to the foetus, suggesting that they should consider postponing travel. Other governments in affected countries, such as Colombia, the Dominican Republic, Puerto Rico, Ecuador, El Salvador and Jamaica, went further, urging women to consider postponing plans for getting pregnant until more was known about the virus and its risks.

One only need take a good look at the flaviviruses, and the arboviruses in general, to see why doctors were alarmed. The most infamous member of the family is the yellow fever virus, from which it derives its name: *flavus* is the Latin for 'yellow'. The yellow refers to the jaundice produced by viral damage to the liver. Yellow fever is one of the most notorious plagues of history. Its prevalence in the tropical and subtropical ecologies of Africa was a contributor, together with malaria and other contagious infections, to the fact that, during the period of European colonial expansion, Africa became known as the 'white man's grave'. Other members of the flaviviruses include dengue fever, which is also known as 'breakbone fever', and Chikungunya, which is one of the so-called 'haemorrhagic fevers'. All three flaviviruses are transmitted by the *Aedes aegipti* mosquito. Other non-flaviviruses transmitted by insects include the West Nile virus, tick-borne encephalitis virus, Japanese B and Murray Valley encephalitis, St Louis encephalitis, and some assorted others.

The arboviruses, including Zika, are relatively small, with virion diameters of 37 to 65 nanometres, with an icosahedral capsid overcoated by an outer lipoprotein membrane. The yellow fever virus was the first virus of humans to be isolated in medical history. Prior to recognition of the mosquito vector, and before the introduction of the vaccine, yellow fever was one of the most lethal infections to afflict our species. Historically it seems likely that it spread from Africa to South America through the slave trade, with the result that it is now endemic in both continents. The host reservoir is limited to primates, including humans. The illness is particularly severe in people unlucky enough to be re-infected with the same illness, but even the primary infection can be life-threatening in children or in those with a reduced immunity, such as diabetics. Alas, as with Zika, our current range of antiviral drugs is relatively ineffective. Globally, in 2013, yellow fever caused a total of approximately 137,000 infections, which resulted in 45,000 deaths, mostly in Africa. Tragically these illnesses and deaths might have been prevented by prior vaccination.

In December 2016 the actor Tony Gardner revealed to the *Times* newspaper reporter Kaya Burgess that he had contracted the Zika virus while filming the BBC series, *Death in Paradise*, in the Caribbean. In that same year an estimated 265 British travellers were said to have been infected with the virus. For virologists, and doctors involved with public health, the mercurial nature of Zika's behaviour, in particular the growing severity of its complications, was presenting them with a new panoply of problems. This resulted in a renewed clamour of health warnings to travellers heading for parts of the world where Zika was now epidemic. This was the situation that faced athletes heading for Brazil in 2016.

But what were the athletes to do?

They had devoted years of their lives to preparing for the Olympics, when they might perhaps have this one chance of winning a medal. At the same time many of the female athletes were of reproductive age and thus needed to be warned that any local pregnancy might result in the disaster of a child being born with microcephaly. The television screens were filled with images of these hapless babies, with the upper halves of their skulls shrunken because of the poorly developed brains. In January the Brazilian authorities mitigated the risk, ordering the inspection of the facilities in advance of the Games, and detailing a small army of field operatives to eliminate any potential breeding grounds for mosquitoes. Daily sweeps were planned to take place during the actual Games, keeping fumigation to precise localities so there would be no ill-effects of the fumigation on athletes and visitors. Meanwhile the health authorities in participating countries, such as the UK, continued to issue up-to-date travel advice, including 'mosquito bite-prevention-strategies'.

How remarkable, in retrospect, that a virus that had started out as an endemic infection in monkeys and chimpanzees in a small region of African rain forests had, over the course of a single century, become a global threat to human health. In the United States more than 5,000 cases of Zika virus infection were reported between 1 January 2015 and 1 March 2017. Most of these were travellers returning from Zika hot spots outside the US but six cases in Texas and some 215 cases in Florida were put down to local infections from mosquito-borne transmission. By August 2016 more than 50 different countries worldwide had experienced some local invasion and spread of Zika virus. The race was now on to stop a Zika virus pandemic. The World Health Organization and the CDC were of the same opinion: priority should be given to developing a preventive vaccine against the virus. By March

2016 various private companies and medical institutions were hard at work to achieve this end. Some adopted an ingenious biological approach.

There is a genus of symbiotic bacterium, called *Wolbachia*, which infects a high proportion of insects, giving rise to an extraordinary interaction with the insect's life cycle. For example, *Wolbachia* selectively infects the insect's gonads, choosing to infect mature ova while ignoring mature sperm. This ensures that infected females pass the germ on to their female offspring. The germ also selectively kills male offspring during larval development. Any infected male offspring that manages to survive to adulthood has its metamorphosis subverted so it develops into an infertile 'pseudo-female'. In some species of insects, such as the *Trichogramma* wasp, the insect's reproduction is diverted to exclusive parthenogenesis – females give rise to exclusively female generations without the role of a male partner. A droll observer might consider it a metaphor for the ultimate female liberation.

Wolbachia does not normally infect *Aedes aegypti* mosquitoes, so there was no easy application of this bizarre symbiological interaction to the Zika problem. But in other insects, concomitant infection with *Wolbachia* was known to reduce the ability of arthropod-spread viruses to reproduce within the insect tissues, thus rendering them less likely to transmit the viruses during the human blood meal. Some researchers had been working on field trials based on infecting *Aedes* mosquitoes with *Wolbachia* before releasing them in large numbers into ecologies where insect-transmitted viruses, such as Zika, are a threat to the human population. In Australia, for example, Professor Scott O'Neill and his team at Monash University had been conducting such field trials of *Wolbachia*-based disease prevention for a decade.

In March 2016 the Australian group was given the go-ahead

by the WHO to trial *Wolbachia*-infested *Aedes aegypti* in Brazil and Colombia in an operation aimed at reducing the threat of Zika virus.

Alas, there would be no opportunity of putting this extraordinary ecological approach to the test, because, in late 2016, the Zika virus changed its behaviour yet again. Out of the blue, the numbers of new cases in the Americas plummeted. At the same time a diminishing infectivity was seen globally, so much so that by November of that year the World Health Organization declared that the Zika virus, though still representing a 'highly significant and long-term problem', no longer represented a global emergency. In Brazil, where some 170,535 cases had been reported in 2016, between January and April of 2017 the numbers of new cases fell by 95 per cent, leading the Brazilian authorities to declare that the national emergency was over. As to the risk of infection in the athletes travelling to take part in the Olympic Games, none contracted the Zika virus during their stay in Brazil. Instead some 7 per cent contracted a more predictable gamut of other insect-borne viruses, including 27 cases of West Nile and Chikungunya viruses and two cases of dengue fever.

Why then had the Zika epidemic petered out?

In fact, the change was probably not brought about by change in the virus alone but by change in the virus–host interaction. We should remind ourselves that viruses are symbionts; their evolution can only be understood through the perspective of the virus–host interaction. In the opinion of Professor David Heymann, Chairman of the WHO emergency committee for the Zika outbreak, the falling numbers of infected were most likely brought about by rising 'herd immunity' among the human population, perhaps because so many people had already been infected. However, we should recall all those earlier changes in behaviour of the virus.

If there is a single characteristic typical of Zika, it is its mercurial unpredictability. Professor Heymann was altogether timely in warning us that Zika had not actually gone away. On the contrary, it had greatly increased its geographic range and it had shown a menacing facility for discovering new modes of transmission.

It would appear prudent to watch and wait and, all the while, remain ever vigilant when it comes to viruses.

14

A Taste for the Liver

Viral hepatitis is one of the most serious contemporary plagues of humanity. The problem is a global one, involving a handful of different viruses. The historical identification of the causative viruses was one of the most fascinating scientific investigations of the second half of the twentieth century, and one that has revolutionised certain aspects of epidemiological medicine and public health. It has also introduced new ways of preparing protective vaccines, including the first application of genetic engineering in their preparation. But what of the viruses themselves and their behaviour once they find the target organ: the liver? This organ is the main biological factory of our bodies. It is extensively involved in the digestion of food, in a comprehensive variety of protein manufacture, for example, the blood-clotting factors, and in the detoxification and removal of potential poisons that enter the bloodstream. It is also heavily involved in the fightback against the blood-borne stages of many of the previously described viral fevers, becoming a site of massive viral replication, but fortunately for us all, paradoxically escaping serious damage. This saving grace is down to the fact that these viruses target the Kupffer cells of the liver, which are an intrinsic part of the reticuloendothelial system, which deals with the immune reaction to foreign invaders. The true 'hepatitis' viruses preferentially target the glandular cells of the liver, called the 'hepatocytes'.

To understand this key observation, we need to understand how the liver works as the largest glandular organ in the body.

Architecturally, the liver works through its unique microscopic structure, which is known as the lobular anatomical architecture. And given the many different functions it serves, it means that the liver is a vital organ, as necessary for life as the heart or the lungs. We are familiar with the term 'cirrhosis of the liver', which implies a process of severe damage to the lobular architecture of the liver through repeated or persisting damage to the hepatocytes. One of the commonest causes of cirrhosis is long-term excessive consumption of alcohol. Contiguous hunks of the lobules are destroyed, followed by extensive scarring, which destroys the fine lobular architecture. This inevitably reduces the efficiency of the many different functions the liver provides. A similar process of cirrhosis can also be caused by persisting viral infection of the liver.

We have already come across viruses capable of infecting the hepatocytes, such as herpesviruses, cytomegalovirus, Epstein-Barr virus and the yellow fever virus. Our modern world is also subject to prevailing endemics of five different groups of viruses, known as hepatitis A, B, C, D and E, all of which target the hepatocytes. These viruses have no kinship with one another and they give rise to differing patterns of disease, making it essential that medical authorities approach each virus in its own light, examining its particular idiosyncrasies of taxonomic classification, structural anatomy, modes of transmission and, most important of all, making use of such knowledge to address specific prevention and therapy.

Hepatitis A, like the poliomyelitis virus, is a picornavirus. We might recall that this implies a very small virus with an RNA-based genome. Under the electron microscope the virion of HAV looks similar to the poliovirus, and both viruses are transmitted by the same faecal-oral route. HAV is tiny, even among the viruses, with a virion diameter of 27 nanometres. Within the family of picornaviruses, it is a species within the genus of the enteroviruses – viruses that spread by the faecal-oral route – with the serotype number enterovirus 72. Unlike other enteroviruses, HAV is extremely difficult to propagate in cell cultures or laboratory animals, a problem that bedevilled early study of the virus. HAV causes 'hepatitis A', commonly referred to as 'infectious hepatitis', which is highly contagious by the faecal-oral route, and commonly infecting children. It has an incubation period of two to six weeks. Despite the fact that HAV is resistant to the acid digestion of the stomach, and despite the fact that the virus replicates within the intestine, it produces no symptoms of gastroenteritis. We might recall a similar situation in relation to the poliomyelitis virus. After replication in the gut, HAV enters the bloodstream, gaining access to the hepatocytes, where it causes the clinical features of hepatitis. Even then the illness may be mild enough to pass unnoticed. When clinical, it starts with malaise, abdominal discomfort and fever, with jaundice becoming obvious within a few days. It rarely causes serious complications and thankfully death is rare. In the scientific jargon the 'virulence' is said to be low.

HAV is excreted in the faeces of infected individuals and survives for long periods in water or wet environments. Hence the disease is more prevalent in countries with inadequate sewage treatment and lack of personal hygiene. Globally the annual infection rate runs to millions. Control measures aimed at prevention of hepatitis A include maintenance of good-quality hygiene and, where authorities feel it necessary, prevention by passive immunisation with human immunoglobulin, or by vaccination with the hepatitis A vaccine. The relatively benign behaviour of hepatitis A virus is in marked contrast to hepatitis B.

Hepatitis B virus, or HBV, is a member of the family of hepatitis DNA viruses, or 'hepadnaviruses'. Like HAV it is extremely difficult to propagate in the laboratory, a problem that delayed the initial discovery of the causative virus. The first clue to its existence came from a chance observation by geneticist Baruch Samuel Blumberg, who found a mysterious antibody in the blood of a haemophiliac patient who had been the recipient of multiple blood transfusions. Blumberg subsequently discovered that the mystery antibody matched an antigen in the blood of an Australian Aboriginal. The antigen turned out to be a piece of the hepatitis B virus. This ultimately led to the recognition of hepatitis B as a life-threatening epidemic disease, leading further to the development of the preventive HBV vaccine. In 1976 it also led to Blumberg sharing the Nobel Prize in Physiology or Medicine with virologist D. Carleton Gadjusek, who had discovered the bizarre cause of Kuru, a disease brought about by cannibalism in a Stone Age tribe in New Guinea. Kuru was caused by the same abnormal transmissible protein, or 'prion', that causes 'mad cow' disease in cattle and Creutzfeldt-Jakob disease in people.

As the family name, hepadnaviruses, suggests, HBV possesses a DNA-based genome, wrapped in an icosahedral capsid, which is further enveloped by a glycoprotein membranous shell. Viral transmission is quite different from HAV in that HBV is not contagious through the faecal-oral route but through contact with blood, or through body fluids, including cervical secretions in females and semen in males. HBV also differs from HAV in having a much longer incubation period, which varies from two to five months after first infection. Once it enters the bloodstream the virus exclusively targets the hepatocytes, where it replicates on a grand scale to return vast numbers of viruses back into the bloodstream. The blood of an untreated individual is so contagious that as little as one ten-thousandth of a millilitre of contaminated blood is capable of transmitting the infection to another human being. This means that even minor abrasions or breaks in the skin or mucosa could enable entry of the virus, facilitating spread through sexual intercourse, particularly among gay males, or through the sharing of needles or syringes among drug addicts.

The liver has extraordinary powers of recovery, including the ability to regenerate itself after major losses. But it also has that unique vulnerability of structure-makes-function that requires the integrity of the lobular anatomical architecture. Damage to this causes cirrhosis and, eventually, life-threatening liver failure. But with hepatitis B, there is an additional risk of liver cell cancer.

According to the World Health Organization, an estimated 257 million people are living with HBV infection today, despite the fact that a preventative vaccine, which gives 95 per cent protection against viral infection, has been widely available since 1982. In 2015, the virus resulted in 887,000 deaths globally, mostly from complications such as cirrhosis and liver cancer. This is a major risk to many people in disparate countries worldwide, but with relatively high rates of infection in the Western Pacific region of south-east Asia and in Africa, in each of which territories an estimated 6 per cent of the adult population is infected. Globally, there is also a connection between HBV and HIV infections, with some 7.4 per cent of HIV-infected people also suffering from concomitant HBV infection.

Unlike HIV, there is no specific cure for HBV infection, although chronic infection with the virus can be mitigated by oral antiviral drugs, which can slow the progression of cirrhosis and reduce the incidence of liver cell cancer.

In the 1970s a third cause of hepatitis was found in patients who were seronegative for both HAV and HBV. Initially dubbed Non-A Non-B hepatitis, the cause is now recognised to be a different virus, known as hepatitis C, or HCV. In fact, HCV is another flavivirus, which is transmitted by contaminated blood. Like its family relative, Zika, it is also capable of crossing the placenta of an infected mother to infect the developing foetus in the womb. Curiously, there appears to be a very low risk of sexual transmission of HCV and it may not be infectious from mother to baby through routes other than the placental transmission. Like other flaviviruses, it has an RNA-based genome, with a relatively small virion of 55–65 nanometres.

Like HAV and HBV, hepatitis C is exhibiting a similar pattern of highly infectious global spread. Some authorities believe it may prove to be even more dangerous in terms of cirrhosis and hepatocellular cancer than HBV. In 2017 the number of new HCV infections in the US alone was found to have tripled over the previous five years, making it the commonest prevailing blood-borne infection in the country. It currently infects some 200,000 residents of the UK, often remaining asymptomatic for decades, but eventually culminating in cirrhosis. There is a peculiar cross-linkage of pathologies that results in the fact that if a sufferer from hepatitis C contracts concomitant HAV or HBV, this can result in a worsening of hepatitis. This is why all sufferers from HCV are advised to get immunised against HAV and HBV if not already immune.

The good news in this somewhat gloom-laden scenario is that HCV levels in the blood can be reduced to undetectable levels by a combination of interferon and antiviral drugs. Indeed, the health authorities in the UK currently believe that direct therapy of this sort may already be improving the mortality statistics relating to HCV in the UK.

Hepatitis D, or HDV, is caused by a tiny agent belonging to the *Deltavirus* genus, which is too small and defective to infect humans by itself. It can only propagate itself in the presence of HBV, with the latter serving as a 'helper virus'. The hepatitis E virus, or HEV, is more prevalent in developing countries, where it produces an illness similar to HAV. But this virus is not exclusive to humans, being infectious to a wide range of animals, including those farmed for their meat. The resulting illness is usually mild and self-limiting but among pregnant women there is a risk of hepatitis that can lead to liver failure. There are four strains of the virus, with strains 1 and 2 limited to Asia and Africa, strain 4 limited to China and strain 3 more globally distributed. Recently some UK newspapers warned of this strain being found in meat imported from Europe, with the warning that it had infected tens of thousands of patients in Britain during 2017. These statistics have not, at the time of writing, been officially confirmed by UK.gov or British Liver Trust websites. However, there does appear to be a slowly rising incidence of HEV, year by year.

All of these liver viruses are examples of what is termed 'emerging infections'. Emerging viral infections are both unwelcome and frightening and they threaten every form of life on Earth. I'm afraid that humans are no exception. Some of the viruses that afflict us are particularly dangerous because they invade and replicate within the landscape of our genome.

15

Warts and All

Prior to his portrait being painted by Sir Peter Lely, Oliver Cromwell instructed his artist: 'Mr Lely, I desire you would use all your skill to paint my picture truly like me and flatter me not at all. But remark all these roughnesses and pimples, warts and everything as you see in me. Otherwise I will never pay a farthing for it.' The world of medicine has learned to pay a similar attention to that same common wart, and all that its presence might imply.

The very word, wart, which derives from the Saxon *warta*, seems appropriately onomatopoeic for the ugliness of this condition. Most of us are familiar with warts erupting through our otherwise smooth expanse of skin, like minuscule cauliflower heads, whether they disfigure our hands or the plantar surfaces of our feet. Some of us have also suffered the indignity on more intimate parts of our anatomy. We are, alas, vulnerably human, driven by human needs and desires. Genital warts, which are usually multiple as well as being highly contagious, afflict the cervix, vulva and vagina in women, and the penis and perianal region of men. Less frequently, but perhaps even more distressing, warts can crop in the mouth and throat as a result of oral sex in both men and women. Given their obvious visible nature, it is hardly surprising that warts were familiar to sufferers and physicians alike since ancient times. The term 'verruca' was coined by a German physician, Daniel Sennert, in 1636, based on the Latin for a small hillock arising out of the surrounding flat plain of skin. The clinical term for genital warts, *Condylomata acuminata*, is derived from the Ancient Greek *condylomata*, which describes a knuckle or a knob, and *accuminata*, which emphasises their proliferative nature – a 'venereal' condition that would have been familiar to Hippocrates.

The common wart, or *Verruca vulgaris*, is transmitted by skin-to-skin contact from an infected person, or through contaminated clothes or other surfaces. In 1907 the first hard evidence of its contagious nature came from the observations of an Italian doctor, Giuseppe Ciuffo, who showed that warts could be transmitted by extracts passed through Chamberland-Pasteur filters. The causative virus was subsequently shown to be the papillomavirus. Unfortunately, difficulties in finding a suitable tissue culture for growing the virus delayed further research into the pathological implications for some 60 years, with a cost in terms of human suffering and premature death that must have been very considerable.

We can't help being somewhat squeamish about diseases that affect the more intimate parts of our anatomies. But doctors are obliged to put aside the societal stigma or mores associated with the sexual act and adopt a dispassionate clinical objectivity. Cancer of the womb, and in particular cancer affecting the uterine cervix, is one of the commonest types of cancer to afflict women. Uterine cancer was also familiar to Hippocrates, but until recently it tended to be diagnosed late in the course of the disease, leading to a disastrously high mortality rate. Another advance in understanding came from a chance observation by a surgeon, Rigoni-Stern, working in Padua in the mid-nineteenth century, who observed that nuns had a similar frequency of deaths from breast cancer as married women, but they had a much lower frequency of uterine cancer. Given that nuns tended to be virgins, this suggested that there might be a significant link between uterine cancer and sexual activity. This suspicion was heightened when epidemiologists observed that cervical cancer was commoner in sex workers. It was also commoner in women whose husbands had a high number of sexual partners, including prostitutes.

These observations led to a growing suspicion that uterine cancer might have a contagious cause.

While the uterus and cervix are hidden from ordinary bedside clinical examination, gynaecologists found ways of performing superficial examination of the cervix through inspection of the vault of the vagina. In 1925 the invention of the colposcope enabled a closer examination of the cervix. Biopsies could now be taken from the surface skin of the cervix, and through passing forceps through the opening, biopsies could also be taken from the endometrial lining of the womb. Further gynaecological ingenuity over succeeding decades led to the development of the 'Pap' smear technique, which ushered in the advance of cervical cytology.

An enterprising gynaecologist took advantage of the fashion, then prevalent in Australia, for would-be brides to be screened for potential future sexual difficulties in advance of the wedding. With due permission this also offered the opportunity of performing cervical cytology in asymptomatic young women. The comparison of cervical smears from virginal girls and a cohort of similar age who were sexually experienced confirmed that sexual activity was indeed linked to abnormal cervical cytology of the sort associated with pre-cancerous change. Further epidemiological study throughout the 1960s and 1970s suggested that cancer of the cervix, vulva and vagina in women, and cancer of the penis in men, and also cancers of the anus, and even some cancers of the mouth and throat in both sexes, were now linked to sexual activity. It all pointed to a transmissible agent. There was a growing consensus that this transmissible agent would turn out to be a virus.

However, there were conflicting views as to which virus was the most likely cause, with the majority opinion deciding on a herpes virus. The herpesviruses, which we examined in an earlier chapter, are a large and varied group that cause a wide variety of diseases in humans, including genital infection, and, most relevantly, several varieties of cancer associated with the Epstein-Barr virus. The focus on herpesviruses seemed eminently logical. But not every virologist was convinced. In 1976, a German virologist, Harald zur Hausen, working at the University of Erlangen-Nürnberg, contradicted the majority viewpoint with a single-page report in the journal *Cancer Research*. Zur Hausen proposed that the cause of cervical cancer was more likely to be the infectious virus that caused genital warts. In his words: 'The condylomata agent has been entirely neglected thus far in all epidemiological and serological studies relating not only to cervical and penile, but also to vulvar and perianal carcinomas. This is particularly unusual in view of the localization of genital warts, their mode of venereal transmission, the number of reports on malignant transition, and the presence of an agent belonging to a well-characterized group of oncogenic DNA viruses.'

Oncogenic means tumour-provoking. The 'well-characterized group of oncogenic DNA viruses' zur Hausen was referring to was none other than the papillomavirus, familiar as the cause of the common wart. Unfortunately, it would take some 30 or so years for zur Hausen's viewpoint to be confirmed. In 2008, his pioneering contribution would be belatedly rewarded when, alongside Françoise Barré-Sinoussi and Luc Montagnier, the discovers of HIV-1 as the cause of AIDS, he shared the Nobel Prize in Physiology or Medicine.

What then are the papillomaviruses? What is it about these viruses that links the cause of the common wart to the menacing potential of life-threatening cancerous change?

As zur Hausen indicated, papillomaviruses have a DNA-based genome. They are modest in size for DNA-based viruses, with an average virion diameter of 55 nanometres and, like many small viruses, they have no protective surface envelope, their genomes merely enclosed in familiar icosahedral capsid. Such details are important when considering viruses since the proteins of the capsid are the first point of contact during the attachment of the virus to the target human cell – and thus the capsid proteins are likely to figure during the recognition of the virus as 'alien' in the ensuing battle with the human immune reaction to the presence of the virus.

Papillomaviruses are spherical, as seen under the electron microscope, resembling minuscule golf balls. They make up a genus within the viral family known as papovaviruses, which includes just one more genus known as the polyomaviruses. It is no accident that the 'oma' in the names of both genera is the same 'oma' in the term 'carcinoma', which is a medical term for cancer. This warns us that both these genera of viruses are known to cause cancers.

We have seen how choosy most viruses are when it comes to host species. That specificity includes the precise targeting of host target cells, and this targeting often involves an interaction between the viral capsid or its enclosing envelope, to a specific chemical receptor on the host target cell. The target cell of the human papillomavirus is the stratified squamous epithelial cell of the human skin. Indeed, the virus is even more specific than that: it can only replicate in a skin cell that is actively 'differentiating' – this means a stratified squamous cell that is itself in the process of replication through the process of mitosis.

This extraordinary specificity explains the earlier difficulty in finding a suitable culture medium for growing the virus, since actively differentiating stratified squamous cells cannot be grown in conventional cell cultures. Moreover, this same specificity is of cardinal importance when it comes to understanding how the warts virus causes cancer. Mitosis is a very complex process, involving the replication of the entire human genome, with its full complement of 46 chromosomes, as readers might recall from their lessons in school biology classes. The very notion that papillomaviruses might invade this process in order to replicate themselves is actually mind-boggling in its implications.

There are 170 strains – the equivalent of species – within the genus of human papillomaviruses, or HPVs, of which some 40 or so strains are capable of spreading through sexual contact. These strains infect the skin cells of the genital areas in infected patients, and sometimes the oral cavity of men and women. A dozen or so of these strains have now been linked to sexually associated cancers, including cancers of the cervix, uterus, vulva, vagina, penis, peri-anal skin and the throat. All these body regions are lined by stratified squamous epithelium. In 2002, when the causative link between papillomaviruses and this range of cancers was recognised, medical epidemiologists estimated that some 561,200 new cancer cases worldwide fitted the picture illustrated above and thus could be attributed to HPV in that single year.

The detailed investigation of how HPV causes cancer is still ongoing. Why should papillomavirus infections cause so little in the way of symptoms in the case of common warts, which, for example in children, very often resolve spontaneously, yet when transmitted sexually give rise to infections that not only persist, but also carry the potential for deadly disease?

In fact, research suggests that most of the papillomavirus infections of the cervix, as with warts on the skin, will also heal without causing significant disease. In such cases, we must assume that the body's own defences curtail the viral presence. But a minority of viruses, identified as specific genotypes, or 'strains', appear more likely to lead to cervical cancer. Additional risk factors for persistent virus infection in males and females include early age of first sexual intercourse, multiple partners, concomitant smoking and poor immune function. The majority of these dangerous strains are acquired through venereal infection, but occasionally a dangerous strain can be spread from mother to baby during pregnancy.

In a bulletin of the World Health Organization, dated a decade or so ago, cervical cancer was still estimated to afflict approximately half a million women per year. Some 80 per cent of these were in developing countries, where impoverished health resources were unlikely to provide a timely cure. In an article in *The Lancet*, published in 2018, HPV infection was implicated in more than 99 per cent of cervical cancer cases, with roughly 70 per cent attributed to two specific strains of the virus, HPV-16 and HPV-18. Today we have a better understanding of how these viruses hijack the cellular DNA of the multiplying skin cells and thus give rise to cancer.

In the actively dividing cells the interfering viral presence within the reproducing genome is a genetic spanner in the works. It provokes mistakes in the copying of the DNA of the host cell from one generation to another. Mistakes in copying DNA are, by definition, mutations, which are inevitably inherited by future generations of the skin cells. Generation after generation, the same process of viral hijacking continues, and so the number of mutations in the dividing squamous cells gradually accumulate. Today we know that cancers of all types arise through gradual accumulation of mutations. Once this accumulation of mutations takes place, it is impossible to reverse it even with all of our innate genomic defences, and despite the prevailing stratagems of modern medicine.

But through understanding how the viruses cause the cancerous disease, we are now in a better position to deal with it. We can treat the cancers when they are diagnosed, by means of surgery, anti-mitotic drugs and other therapies, such as focused radiotherapy. Such therapies are more likely to be successful if and when the cancers are caught early – this depends upon well-organised and efficient screening programmes. Better still is to prevent the cancers from arising in the first place. In part this is helped by education aimed at risk reduction, which needs to be addressed as early as possible to our young people. As recently as 2005, HPV still caused an estimated 260,000 deaths worldwide through cervical cancer alone. Although the bulk of such deaths were confined to developing countries, as recently as 2017 the CDC estimated that some 79 million Americans were still infected with HPV, with some 14 million new infections arising annually. The authorities also estimated that some 4,210 American women would die from cervical cancer that year. While emphasis was placed on the importance of diagnosis for effective therapy of all HPV cancers, there was also a cogent need for vaccination of the young against infection with the known high-risk strains of papillomavirus.

Vaccines prepared from virus capsid proteins have now been available for more than a decade. In 2008 the UK initiated school-based vaccination of 12 to 13-year-old girls, and by 2012–2014, more than 86 per cent of the population had received a full course of vaccination. By 2017, a study of the prevalence of HPV infection in young women in Scotland showed that it had fallen by 90 per cent following a successful vaccination campaign. Health Protection Scotland had every reason to anticipate that this would result in a major drop in cervical cancer in future years. More widely in the UK demographic studies reported substantially lower levels of vaccine uptake in girls from ethnic minority backgrounds, most markedly in those of Asian origins. The same ethnic minority disparity was reported in cervical screening attendance. The authors concluded that the introduction of HPV vaccination in England would most likely widen a pre-existing disparity in the incidence of HPV-related cancer by ethnicity.

In 2014 the FDA approved the vaccine, Gardasil, as a preventive therapy for both males and females in America, with specific guidelines in terms of age, level of immunocompromisation and sexual proclivities. Despite relatively few adverse effects, vaccination uptake varied between different states, but to date the uptake would appear to be rising states-wide. The efficacy of the vaccine may have been undermined by limited uptake in some countries, including the Republic of Ireland, where, despite Ireland reportedly having one of the highest rates of cervical cancer in Europe, half the eligible young women refused the cervical cancer vaccine between 2016 and 2017. This is a cause for concern for the Irish Health Authorities, who are reinvigorating their vaccination campaign, so that the local medical authorities will put a renewed effort into protecting the future health and lives of Irish women.

A secondary, but perhaps equally important, aspect of HPV vaccination is the problem that vaccination aimed exclusively at women will not remove the reservoir of the causative virus in the population at large, and equally important, the associated risk of cancer in men. We live in enlightened times when sexually transmitted viruses are a major risk to all manner of relationships. As already recommended in America, it makes clinical sense that the viral source needs to be eliminated with equal attention to education and organised programmes of early vaccination in all of our younger generations.

16

Lilliputian Giants

In previous chapters we have examined the role of viruses in a wide variety of human diseases. In doing so we have gathered some valuable insights into the medley of infections that result from our sharing the planet with them. One such insight, and one I make no apology for repeatedly stressing, is the fact that viruses are apt to surprise us. This is a facet that impressed me when, as a young medical student, I first gazed at the strange beauty of the bacteriophage viruses I was researching under the powerful magnification of the electron microscope. But never in my wildest dreams did I imagine that one day we would discover what proved to be relative giants in the Lilliputian world of the viruses.

When, in 1992, the Mimivirus was discovered inside the body of an amoeba, it provoked a mixture of shock and disbelief even among experienced microbiologists. Mimivirus was discovered by accident during research into the cause of community-acquired pneumonia, a condition more popularly known as Legionnaire's Disease. Its discoverers were a group of microbiologists from Marseille, in France, and Leeds, in England, who were looking for new strains of the causative bacterium. When they came across a microbe, plucked from a cooling tower in the industrial north of England city of Bradford – a bug as large as a bacterium and which stained like a bacterium with the commonly employed Gram stain used in classifying bacteria – they assumed that they had discovered a new species of bacterium. They named it after the city in which it was discovered, the Bradford bacterium, or in scientific jargon, the '*Bradfordcoccus*'. But when they examined their discovery in closer detail, they were astonished to discover that it wasn't a bacterium at all but rather a virus – albeit a very strange virus indeed. For a start it was truly enormous in viral terms, with a virion capsid diameter of more than 400 nanometres. This Lilliputian giant would never pass through a Chamberland-Pasteur filter. In time geneticists would discover that it also had a much more complex genome than would normally be found in a virus, bigger even than that found in some of the smaller bacteria.

Because it mimicked a bacterium, the new discovery was renamed the 'Mimivirus'. Meanwhile its discovery provoked a dilemma of contradictory opinions among microbiologists. Was it a singular anomaly – or did it herald an exciting new branch of microbiology? It was inevitable that other microbiologists would search for similar giant viruses in aqueous environments throughout the world. Soon Mimivirus proved to be the harbinger of a growing plethora of giant viruses, now classed as 'Megaviruses', including *Megavirus chilensis*, *Pandoravirus salinas* and *Cafeteria roenbergensis* – the latter isolated from seawater collected from the Gulf of Mexico, and named after a bacteria-gobbling marine protist that was infected by the virus.

Infection is a very approximate term to use in this respect since, as far as we currently know, none of the giant viruses appear to parasitise their hosts in the sense of causing them any illness or harm. *Cafeteria roenbergensis* virus is distantly related to Mimivirus but its genome has far more protein translational sequences.

Protists are single-celled nucleated organisms. Predation by 'protist grazers' is integral to carbon recycling in marine and freshwater ecologies, suggesting the possibility of a symbiotic interaction between these oceanic protists and their giant viruses. In the words of marine microbiologist Curtis Suttle, 'We know next to nothing about the role viruses play in this system . . . but there's little doubt that this virus is just one representative from a major group of largely unknown but ecologically important marine giant viruses.'

Normally we think of viruses as having very simple genomes since they rely on the genetic machinery of their hosts for their replication and biological life cycle. But these viral giants, containing as many as 911 protein-coding and protein translational genes, raised further existential questions about the origins and subsequent evolution of viruses. Two French microbiologists, Jean-Michel Claverie and Chantal Abergel, wondered if the discovery of the Megaviruses challenged the very definition of viruses, the diversity of forms they adopted and how viruses might have evolved in the first place. Another French microbiologist, Patrick Forterre, highlighted this same iconoclastic dilemma in a paper, 'Giant Viruses: Conflicts in Revisiting the Virus Concept'. He re-examined some of the ideas about the origin of viruses that had prevailed for some 50 years or so, highlighting how various authors were now interpreting the significance of the newly discovered giant viruses according to their prefixed prejudices as to viral origins, thus failing to generate a consensus across the wider world of microbiology and indeed the world of biology as a whole. The discovery of the Lilliputian giants was certainly causing waves.

One astonishing Pandoravirus was found to inhabit an amoeba which was discovered in the contact lens of a woman in Germany. Within the virus the reporting biologists discovered a second extraordinary resident, a 'virophage' – a much smaller virus that was parasitising the Pandoravirus. The virophage was named 'Sputnik' after the pioneering Russian spacecraft.

The study of the giant viruses continues to date. They are now seen to consist of an evolutionary order containing a number of different families. Some of these, such as tupanviruses – named after a thunder god of native people in Brazil – are more than a micron in diameter. Another family, the klosneuviruses – found in the wastewater of a treatment plant in Klosterneuburg, in Austria – were found to contain an even more extensive protein-making apparatus. In 2017 a group of virologists, who had developed a bioinformatic shotgun-sequencing technique for screening for the presence of giant viruses in different ecologies, called the 'Giant Virus Finder', astonished biologists even further when they reported an abundance of giant viruses in the arid valleys of Antarctica. The same group went on to extend their searches to numerous hot and cold desert soil samples as well as tundra and forest soils, before drawing the conclusion: 'Giant viruses could be frequent not only in aqueous habitats, but in a wide spectrum of soils on our planet.'

Some microbiologists were of the opinion that the Megaviruses blurred the long-accepted divisions between viruses and cellular life. Some went so far as to propose that they derived from, or perhaps even defined, a fourth domain of cellular life. But detailed genetic study of the three groups, the Mimiviruses, Pithoviruses and Pandoraviruses, showed that each had its evolutionary origins in smaller, well-defined families of DNA-based viruses. This suggested that the giants had come about through ballooning of the viral genomes through the acquisition of numerous genes and genetic sequences from their hosts. While this disappointed people who saw the Megaviruses as a fourth domain of cellular life, it confirmed the earlier suspicion that the relationship between the giant viruses and their hosts was a closely integrated genetic symbiosis.

Today biologists continue to make novel discoveries about the true nature and fundamental role of viruses, some of which we shall consider in the following chapters. If nothing else, it suggests it is high time that we cleared our minds of outdated prejudices about the nature of viruses, re-examining some vitally important questions from the perspective of modern evolutionary biology. Perhaps we should start with the obvious basic question:

What are viruses?

Viruses were formerly defined as 'genetic parasites', but as we explore viruses further this definition will be seen to be far too narrow to accommodate the wide range of interactive relationships that viruses actually share with their hosts. In what may be a helpful reappraisal, the French microbiologists Forterre and Prangishvili have suggested, if we make the step to recognising the fact that viruses are indeed distinct organisms, we should further define them as 'capsid-encoding' organisms as opposed to the 'ribosomal-encoding' organisms of cellular life forms. This would seem a reasonable initial step to make. What is becoming altogether obvious is that, in order to truly understand viruses and their more comprehensive role in the origins of life and its subsequent diversity, we need to examine viruses at the fundamental level of their evolution. Where best to begin such an examination than with the very origins of evolutionary biology with its founding figure: Charles Darwin.

Of course, Darwin had no notion of the existence of viruses – and no more could he have been expected to know anything about modern genetics or genomics, since DNA and RNA were unknown in his lifetime. It is all the more remarkable that when

Darwin proposed his theory of evolution by means of natural selection, he knew that it would only work if there was some system of heredity that could be passed down from parents to children – in his day people called this 'pedigree'. He was also prophetic in grasping that this heredity must also be capable of being altered. Nature could only 'select' between competing individuals, or populations, where it was presented with a range of hereditary variations to choose from. Today we know that heredity involves the inheritance of specific packets of DNA-based information that we call 'genes'. In sexually reproducing organisms, such as animals and plants, this becomes even more complex because it involves the mixing of the heredity of the two parents during the formation of the germ cells, the ovum and sperm. This jumbling up of the genes from both parents is called 'homologous sexual recombination' and it explains why, other than in identical twins, siblings are different from one another.

In Darwin's time naturalists assumed that sexual mixing was somewhat akin to the blending of liquids. They also assumed that the 'variation' resulting from sexual mixing within a species would be enough, over lengthy periods of time, to give rise to the origin of new species. But by the early years of the twentieth century, and with the growing understanding of genes and genetics, biologists realised that the amount of genetic variation within a single species was insufficient to generate a new species from the old, no matter how much sexual recombination took place. The evolution of new species demanded more powerful mechanisms for changing heredity than mere homologous sexual recombination – it required mechanisms powerful enough to make possible, albeit over the billions of years the Earth has existed, the evolution of the rich diversity of life we find in our world today.

If we examine evolutionary change from this genetic perspective,

THIS MAKES NO SENSE

it is obvious that evolution does not begin with natural selection. It must begin with genetic change arising in individuals, which gives these individuals some advantage in terms of survival over other members of the same species. As Darwin realised, this genetic change will indeed be hereditary. The hereditary nature of such change makes it possible for the advantage to spread first to the extended family group, and further, through the continuing advantage for survival, to become incorporated into the local population and then further to the evolving species gene pool. At every step towards change in the species gene pool, Darwinian natural selection will be operating, just as he envisaged it all those years ago. It is this process of genetic or genomic change being 'selected for' by natural selection, again and again, as it works its way from individual to family to species gene pool, that drives evolutionary change.

Today we know of at least four definable and genetically demonstrable mechanisms capable of changing heredity in this way. These include mutation, change in epigenetic inheritance systems, genetic symbiosis – in evolutionary terms this is called symbiogenesis – and genetic change through hybrid sexual crossing, or 'hybridogenesis'. All four can be readily remembered by the acronym, MESH.

We have encountered mutation already in relation to the influenza virus. It is defined as errors in copying DNA during the formation of offspring, in the case of asexual reproduction, and the formation of the germ cells during the process called 'meiosis' in sexually reproductive life forms. Similar mutations can also take place in the non-germ cells, known as somatic cells, during somatic cell mitosis. Somatic cell mutation plays no part in heredity, since it does not affect the germ line. But, as we saw in the papillomavirus chapter, repeated somatic cell mutations is a key process underlying the pathogenesis of cancer.

Epigenetic inheritance systems are a less familiar, yet uniquely important, source of hereditary variation, and they take a bit more explaining. In essence, epigenetics can be conveniently understood as a group of distinct mechanisms contained within the genome that control the expression of genes. These epigenetic mechanisms play a vital role in the development of the human embryo in the mother's womb, determining the differentiation of the single fertilised cell into all the different tissues and organs of the body, and they continue to play a vital role in our normal physiology throughout all of our lives. Disturbances of the epigenetic control mechanisms can cause birth defects and hereditary diseases. I shall have more to say about the mechanics of symbiogenesis in the following chapter and so will defer this for the moment to move on to the 'H' of the MESH acronym – hybridogenesis.

We have also encountered an example of hybrid evolution in relation to the origin of pandemic influenza viruses, but viral hybridisation is quite different from what is meant by hybridisation in sexually reproducing animals and plants, where it refers to evolution through sexual crossing of two different, but usually closely related, species. Those who are not familiar with evolutionary biology might imagine that if two species are closely related, for example, if they have evolved from a single ancestral species perhaps some half a million years earlier, there will be little genetic difference between them. This is quite mistaken; there will be many genetic changes arising through various processes, including mutations, after the species have been evolving over this time period. A merger of the now-dissimilar genomes, which is what results from such a hybrid cross, will create a big jump in genetic diversity in the offspring.

Past generations of geneticists thought hybrid crosses were unlikely in animals, especially mammals, because they thought it would result in an awkward doubling of the chromosomal content of the offspring cells – a process called polyploidy. But today we realise that, provided the hybrid parents are not too genetically dissimilar, the hybrid offspring will have the normal chromosomal complement, a condition known as 'homoploid hybridisation'. In recent years geneticists have discovered that the modern Eurasian human genome is the result of hybrid crosses with other closely related human species, such as Neanderthals and Denisovans. Of critical interest to medical geneticists is the fact that the processes that provide the hereditary change necessary for evolution to work are the same processes that give rise to the genetic component of hereditary and acquired human diseases.

What then does MESH have to do with viruses?

Viruses evolve using very similar evolutionary mechanisms to cellular organisms. Indeed, they evolve by orders of magnitude faster than cellular organisms. And it is in the nature of viruses, in their symbiotic interactions within that vitally important genomic landscape of their hosts, that viruses can sometimes become involved with the MESH system of hereditary change, so altering the evolution of their hosts. This is essentially a symbio-genetic pattern of evolution. We shall return to this remarkable aspect of viruses shortly, but first we need to further clarify the quintessential nature of viruses.

17

Are Viruses Alive?

In 2002 Eckard Wimmer, a professor in the Department of Molecular Genetics and Microbiology at Stony Brook, New York, reconstructed the poliovirus from mail-order components back in his lab. This experiment provoked a mixture of interest and notoriety. What Wimmer and his co-workers wanted to do, amongst other things, was to make a conceptual, even philosophical, point: if you know the genetic 'formula' of a virus, you can reconstruct it. Indeed, they went so far as to write out a purported chemical formula for the poliovirus as follows:

$$C_{332,652}H_{492,388}N_{98,245}O_{131,196}P_{7,501}S_{2,340}$$

Of course, a virus is not a simple chemical compound that can be readily constructed from a list of atoms. There must be biological and evolutionary significance in the projection, and moreover a combination of forces, whether chemical, or, at nucleotide level, coding, that would give meaning to what would otherwise seem more like a meaningless sequence of letters and numerals. Thus, while it might appear that Professor Wimmer is promoting the idea of the poliovirus as an inert chemical, this is not really his thinking at all. When I asked him if he believed that viruses were alive or dead, his response was an enigmatic, 'Yes'.

You have to think about it for a moment or two to take on board his quixotic sense of humour.

In 2009 the microbiologists Moreira and López-Garcia put forward a more brutal rebuttal of the notion of viruses as living beings under the heading: 'Ten reasons to exclude viruses from the tree of life'. For the purpose of even-handedness and clarity, I have gathered their arguments into an easily assimilated form:

- Viruses, being genetic parasites of living organisms, could not have come into existence until the cellular forms, the prokaryotes (Eubacteria and Archaea), had evolved.
- As obligate parasites, and thus being incapable of independent cellular metabolism outside their hosts, viruses are not life forms.
- Viruses do not self-replicate.
- Viruses do not evolve through their own mechanisms. They can only evolve through mechanisms borrowed from their cellular hosts.
- Viruses get new genes by 'pick-pocketing' genes from their hosts.
- Some of the most important families of viruses originated as mere genetic offshoots of their host genomes.
- From the above, no meaningful evolutionary tree of life (phylogeny) can be depicted for viruses.
- Viruses are not cellular and, since life can only be defined from a cellular perspective, they should be dismissed from any consideration as life forms.

Let us concede that this is a well-thought-out argument. It is an argument that I profoundly disagree with, but I am obliged to do so on the basis of equally clear and fact-based scientific principles.

Where to begin? Perhaps I should start with what might seem a surprising degree of agreement with several of these bullet points. I agree that viruses are not cellular life forms. I also agree that this excludes viruses from what is currently referred to as 'the tree of life'. However, this is not yielding to their argument but rather my insisting on a balanced perspective. Since the tree of life that Moreira and López-Garcia are referring to was specifically defined to fit an exclusively cellular perspective, it inevitably excludes viruses. I propose that viruses have many properties specific to life in spite of the fact they cannot be assessed from a cellular perspective.

A virion may appear inert outside its host, but when it enters the host cell it has many qualities that we would attribute to life. It struggles for its own existence in conflict with the hostile world of the host's defences, and having survived this battle it continues the struggle to replicate itself, albeit taking advantage of the host's cellular and genetic physiology to do so. Thus I agree with Moreira and López-Garcia that this implies that a virus cannot complete its life cycle without the presence of its host and partner. But the fact that an organism is dependent on a partner in life in order to reproduce itself does not exclude it from considerations of life. Some bacteria, such as *Rickettsia prowazekii*, the cause of epidemic typhus, are obliged to live within the cytoplasm of their hosts and are dependent on their hosts for survival. If we look to wider associations and dependencies, do not humming birds and their symbiotic floral partners depend on one another for food and pollination? Do not bees depend on the nectar extracted from flowers? Indeed, do the nectar-producing flowers not rely on the bees for pollination? Do not we humans depend on photosynthetic organisms to make the oxygen essential for us to breathe, and for plants, and a medley of other organisms, to make the essential amino acids, essential vitamins, essential fats and other nutrients that enable us to live from day to day? In nature, dependence on another organism for life's essentials is not unusual: it is the quintessential norm for the vast majority of life forms on our planet. Apart from the relatively unusual bacteria known as autotrophs, which can take what they need for existence from inanimate sources, every other species of life on Earth is dependent on the presence of other life forms for its survival.

As to their suggestion that viruses could not have come into existence until there were cellular forms for them to parasitise, let us examine the evidence. Later in this book I shall present an argument to suggest that RNA viruses originated in the RNA world, long before cellular organisms appeared. There was an earlier theory that proposed that virus evolution began as genetic offshoots of their pre-evolved cellular hosts. This is also most unlikely since the majority of core viral genes are absent from the genomes of cellular organisms. This is not to say that some families of viruses did not acquire some host genes at some stages in their evolution – but even this, as we shall duly discover, is a feature of the evolution of all life forms and thus hardly disqualifies viruses from being living. In subsequent chapters we shall discover that viruses have contributed in vitally important ways to the evolving genomes of their hosts, and to the ecological cycles of host survival and death. Genetic exchange and interplay have always gone both ways. Indeed, given the increasing knowledge of genomes, and their capacity for horizontal genetic mobility, the situation is far more complex than Moreira and López-Garcia put forward. As to the dependence of viruses on the genetic and associated biochemistry of their hosts, this is normal in a symbiotic relationship at genetic level. All genetic symbioses, and there are vast numbers of them throughout the tree of life, involve such interactions.

Viruses have definite life cycles, as any virologist, or epidemiologist, would recognise. They are 'born' within the environment of the host target cell. They target specific cells in specific hosts, which should be regarded as the normal ecology of viruses. Here they show recognisable patterns of behaviour, pathology and evolutionary development of the sort that we would expect of living organisms. They replicate following the coded instruction of the virus's own genome, albeit taking advantage of some elements of the genetic or translational machinery of their cellular hosts to do so. The daughter virions have evolved a form capable of abandoning their hosts and moving out into the ecology, there to survive with pre-evolved strategies aimed at discovering new hosts. Contrary to what some imagine to be an inert chemical form, the virion stage is more akin to a seed, which makes use of host locomotion and various behavioural patterns to spread their remit wide and far. Like a seed, it only develops into its fully developed potential when it enters the 'soil' of its target cell. Like other life forms, viruses can die. They are killed by the action of viricidal drugs and they die naturally in nature through exposure to a non-supportive environment or to a wide range of different physical insults.

Rather than the former over-simplistic notion of predator and prey, there is growing evidence that viruses and cellular life have been intrinsically interdependent in their evolution and life histories. Today we have come to realise that viruses and the three domains of cellular life have been entwined in a complex labyrinth of formative evolutionary interactions since the dawn of evolution. The last two decades have seen a sea-change in our knowledge and understanding of viruses that lays bare the inadequacies of overly simplistic definitions of what viruses are and how they evolve. Let us now address the outdated notion of viruses as 'genetic parasites'.

Most scientific definitions of parasitism imply an organism that lives in or about another organism at its detrimental expense. Today we can see that this is far too restrictive a definition to encompass the wide range of interactive roles that viruses play in relation to their hosts. A more meaningful and inclusive definition of viruses, and their complex inter-relationships with their hosts, would employ the concept of 'symbiosis'. A symbiotic relationship between virus and host would include parasitism, where the virus benefits from the relationship at the expense of the host, commensalism, where virus and host co-exist with no detriment to the host, and mutualism, where both the virus and host benefit from the relationship. Thus we might more usefully replace the outdated definition of viruses as 'genetic parasite' with the more accurate and inclusive definition of 'obligate symbiont'. The group of bacteria referred to above, the *Rickettsia*, are also obligate symbionts of their hosts. To dismiss viruses because of their dependency on the host cell for vital biological functions, meanwhile accepting bacteria that exhibit a similar dependency, is logically inconsistent.

An earlier generation of evolutionary biologists, the so-called neo-Darwinian school of thinking, emphasised selfish competition as the drive to evolution. While selfish competition is indeed a powerful evolutionary drive, it is no longer seen as the exclusive avenue to evolutionary successes. Life is not only a bloody confrontation: it also depends on myriad living interactions, from the day-to-day struggle for existence at individual organismal level to the great cycles of nature, including the water, oxygen and carbon cycles. To put it in stark terms, if the humble bacteria found in soil were to disappear, or if the insects were to become extinct, all life as we know it would cease. What role then in the complex vortices of nature are viruses really playing? Could it be that they played a key role in the origins and diversity of life on Earth from the very beginning? In fact, we shall discover in later chapters that viruses are indeed playing a vital role in the ecological cycles on which all of life depends.

We can grasp that viruses are very different entities ƒo cellular organisms. Many, as we have already seen, have genomes made up of RNA rather than DNA, which is unknown in cellular life. All cellular organisms are organised within a compartment enclosed by a cell wall. They possess ribosomal machinery capable of translating genes to proteins. Viruses are not cellular: their genomes are enclosed in capsids, following a polyhedral or cylindrical symmetry, and they do not possess ribosomes. We recall the French virologists Forterre and Prangishvili suggesting that we distinguish viruses from cellular organisms by classifying viruses as capsid-encoding as opposed to the ribosomal-encoding organisms. Let us now pass on to another of the questions raised by Moreira's and López-Garcia's claims.

Do viruses evolve by pick-pocketing genes from their hosts?

It was unfortunate that Moreira's and López-Garcia's dismissal of viruses from the tree of life appears to have been prompted by the claims that the Megaviruses were a new domain of life. I agree with them that the evidence points to these viruses expanding their genomes through capturing a vast array of genetic information from their protist hosts. My argument for the recognition of viruses as organisms does not depend on this viral group. If we look at the field of virology in general, there is abundant evidence to confirm that viruses evolve by means of the normal mechanisms of evolution; they are not dependent on pick-pocketing genes from their hosts. On the contrary, apart from the example of the Megaviruses, the direction of borrowing has been largely in the opposite direction, from virus to host.

How then can we redefine viruses to take into account the

modern perception of their biology and evolution? It is on that basis I would offer a new definition as follows: 'Viruses are non-cellular, capsid-encoding, obligate symbionts.'

This definition is congruous with the fact that close examination of the viral genomes in the modern databases reveals that only a minority of viral genes in nature have come from genetic transfer from host genomes. For example, the genes coding for proteins involved in viral replication are shared by RNA- and DNA-based viruses but are not found in cellular life forms. The same goes for the genes encoding the capsid proteins of icosahedral DNA- and RNA-based viruses. And while biologists in the past took the view that large groups of viruses, such as retroviruses and bacteriophages, came into being as offshoots of their prokaryotic hosts, today the evidence suggests otherwise. Retroviruses and bacteriophages do not derive from host genome, rather they have their own evolutionary lineages, much as any other organismal groups.

People who view viruses only as chemical assemblages, or in isolation without taking into account their symbiotic interaction with their hosts, miss the vitally important point that viruses have arrived on the scene through a vast, and exceedingly complex, evolutionary trajectory, much as we have ourselves. A virus may appear inert outside its host but when it enters the host cell, it metamorphoses into the true biological nature of its being.

It is clearly important that we take the trouble to understand viruses. We have seen how important this is to medicine, veterinary medicine and agriculture, but there is another, equally profound, reason why we should take the trouble to understand viruses. The fact that viruses replicate within the landscape of the host genome confers on them not only the potential for dangerous diseases, but also a unique potential for viruses to change the genome, and thus the evolutionary trajectory, of their hosts. Recent researches have made it increasingly evident that viruses have played an important role in the evolution of life, from its very beginnings on Earth to the magnificent diversity we see today. To understand this role we need to open our vision to a broader perspective of viruses and their true symbiological potential in nature. We might start with a study of viruses in the world of entomology that became a terrifying inspiration . . .

18

Inspiring Terror – and Delight

Perched atop the body of the yellow and green caterpillar of the tobacco hornworm moth, *Manduca sexta*, the parasitic wasp, *Cotesia congregata*, is poised with surgeon-like precision, her syringe-like ovipositor in the act of penetrating the unfortunate creature's skin. When illustrating the scene to my audiences during lectures, I have to highlight the tiny dark-bodied wasp against the relatively vast prey with my laser pointer. This might appear a simple if somewhat blood-curdling act of parasitism, in which the wasp is injecting its eggs into the living tissues of the caterpillar, but there is nothing quite so simple going on here. Had the wasp's eggs been injected alone they would not survive the attack of the caterpillar's immune system, which would destroy them before they could hatch into larvae. But the eggs are accompanied by a subtle if deadly symbiont, in the form of a polydnavirus. Once within the living tissues of the caterpillar, the virus prevents the caterpillar's cellular immune system from attacking the injected eggs. It also prevents the caterpillar from metamorphosing into the adult moth, thus preserving the caterpillar stage as a brood chamber for incubation of the eggs. The same virus then manipulates the caterpillar's biochemical metabolism to induce it to produce nutrients to feed the developing wasp larvae that hatch from the eggs inside the hijacked caterpillar's body, from which the new generation of wasps will duly emerge from what is now an abandoned empty husk.

There are two families of parasitic wasps in nature, the braconids and the ichneumonids, which have entered into this aggressive symbiosis with the polydnaviruses. The wasps are thought to number many tens of thousands of different species – some believe they may even run to hundreds of thousands. The seeming cruelty of the parasitism caused Charles Darwin to exclaim in one of his letters to the botanist Asa Gray: 'I cannot persuade myself that a beneficent and omnipotent God would have designedly created the *Ichneumonidae* with the express intention of their feeding within the living bodies of caterpillars . . .' The same brutal life cycle of the parasitic wasps may have been the inspiration for the terrifying science-fiction film, *Alien*. Today those tens, possibly hundreds of thousands, of species of parasitic wasps form symbiotic unions with similarly vast numbers of species of polydnaviruses to create one of the most successful evolutionary strategies in the entire world of entomology.

Cruel as it would appear to our human sensibilities, some biologists view its operation as generally favourable to the balances of nature, since the wasps contribute a kind of biological pest control on what might otherwise result in teeming numbers of unwelcome insects that can defoliate bushes and trees in profligate manner.

The same parasitic dexterity and cunning applies to a great variety of examples of insect prey, where the wasps are seen to variously attack all stages of development, from eggs to larvae to pupae, and even to the fully-grown adults. For example, *Pompilidae* wasps choose to prey upon spiders. In the words of an expert: 'These spiders are quick and dangerous, often as large as the wasp itself, but the wasp is quicker, swiftly stinging her prey to immobilise it before implanting its eggs.' On Costa Rica another variety of wasp has developed a taste for the orb weaver, *Leucauge argyra*. This spider is considered a formidable predator, but it is reduced to a source of meat by this symbiotic duality of wasp and virus. After stinging, and immobilising the prey, a white grub fastens to the spider's abdomen, and as the grub gnaws and grows, so does the unfortunate spider shrink.

When, and indeed how, could such a complex and diverse symbiotic partnership of insect and virus have possibly evolved?

When geneticists examined the genomes of the polydnaviruses, they discovered a conserved gene family that was present in all of them. This suggested that the vast diversity of present-day symbioses began with a single partnership. The geneticists estimated the likely timing of this primal union at roughly 74 million years ago. We might react with some scepticism to such complexity arising from a single symbiogenetic event at such a great distance in the past, but it is no more astonishing than the fact that our human mitochondria – the so-called power packs within the cytoplasm of our cells that enable our cells to breathe oxygen – are descended from a genetic symbiosis between an oxygen-breathing bacterium and a protist ancestor estimated to have taken place some two billion years ago. That single genetic symbiosis subsequently gave rise to all the oxygen-breathing live forms on Earth today, including all of the animals, plants, fungi and the oxygen-breathing single-celled nucleated organisms we call protists.

The genome of *Cotesia congregata*'s polydnavirus was sequenced in 2004, when it was found to comprise 30 circular chunks of double-stranded DNA. Indeed, the polydnaviruses are unique among viruses in having their genomes broken up in this way into multiple discrete sections, hence the name, poly-dna-viruses. What then do we really know about the polydnaviruses and their aggressive symbiosis with the parasitic wasps? How, for example, do the eggs come to be coated with virus at the key moment when the wasp is poised with her ovipositor plunged deep into the flesh of the caterpillar? The answer is that in this particular species of wasp the viral genome has become integrated into the wasp's own genome.

Not all polydnaviruses are integrated in this way. In some partnerships, the virus simply infects the tissues surrounding the wasp ovaries and the eggs are conveniently coated in viruses as they emerge from the ovaries. But a great many wasps have taken the symbiosis into this deeper incorporation of the viral genome into the nucleus of the wasp, to form a permanent symbiotic fusion of two entirely different genetic lineages – a so-called 'holobiontic genome'.

Evolutionary geneticists have painstakingly explored just how the holobiontic union of host and virus genomes actually works. In a series of bewilderingly complex steps, the viral genome is first put into storage, using a 'virion packaging system', within the nuclei of specific wasp cells, known as the calyx cells, which are located within the female wasp's reproductive tract. From this point, the mechanics of new virus reproduction vary somewhat between the two wasp families. In braconids the virions are released by death and rupture of the calyx cells, while in ichneumonids the virions are released by viral budding through the calyx cell wall. Whatever the complex genetic mechanics, the end result is that virions coat the eggs while still within the female reproductive tract. Thus when the wasps inject their eggs into the caterpillar prey, they are accompanied by polydnaviruses.

The complexity of such a virus–insect interaction would appear as astonishing as it is fiendishly effective. But life in its very essence is quintessentially interactive. As evidenced in the previous chapter, this interactivity is apparent even on a global scale, where key elements are recycled over and over through time. Strange as it might seem, even death involves such elemental recycling, when what was formerly a living body, with all of its complexities of being, is broken down into its component chemicals, these in turn feeding the nutritional needs of the myriad different organisms at the base of the ecological food chains in soil, or the oceans.

One specific example of vital ecological recycling is nitrogen. Elemental nitrogen is, of course, a gas that makes up some 78 per cent of the atmosphere we breathe. It is a key component of amino acids, proteins and DNA. In plants it is essential in the manufacture of chlorophyll, which is integral to photosynthesis. But for this to happen, the inert nitrogen in the air must first be converted into more complex nitrogenous chemical molecules. This in turn is dependent on bacteria, known as rhizobia, which are naturally present in soil. Rhizobia are highly mobile, through the use of their flagella, and they are attracted to chemicals called flavinoids that are released by the roots of legumes. Invading the finest legume roots, called root hairs, they provoke the formation of root nodules. Here the bacteria take in gaseous nitrogen out of the atmosphere and fix it with hydrogen to form ammonium. This is a first step in the formation of more complex nitrogenous chemicals, which are exported to the plant. In return the plant supplies the rhizobia with oxygen, for cellular respiration, and carbohydrates derived from photosynthesis, which provide the rhizobia with their basic energy requirements. This is how the rhizobia–legume symbiosis works as part of the globally important nitrogen cycle.

In nature, however, many strains of soil-based rhizobia lack the fixation and nodulation genes necessary for the nitrogen cycle to work. Back in 1986, a horticultural experiment carried out in a remote field in New Zealand looked for an explanation for how nature remedies this situation. The botanists were conducting experiments on the growth of *Lotus corniculatus*, a flowering plant of the pea family found throughout Eurasia and North Africa. They confirmed to begin with that, although the soil contained plenty of indigenous rhizobial bacteria, none of these were capable of forming nodules on the roots of the plants. But when they planted seeds of *Lotus corniculatus* coated with a single strain of rhizobium, known as *Mesorhizobium loti*, the problem in the local soil was cured. This posed a new question: how exactly had this cure come about?

In fact, through further genetic experiment, they discovered that the local non-fixing rhizobia had been transformed by the transfer of a six-gene 'symbiosis island' from the transplanted *Mesorhizobium loti*. But there was yet another revelation within the symbiotic solution. The symbiosis island could only be transferred from a fixing to non-fixing strain of rhizobium through the contribution of an 'integrase' gene that was intrinsic to the symbiosis island. Bacteria, such as rhizobia, do not possess an integrase gene, but a bacteriophage virus, called a P4 bacteriophage, does. The viral integrase was the smoking gun that pointed to a key additional bacteriophage–rhizobial genetic symbiosis that must have played an earlier, subtle role, in the evolution of the symbiosis island.

The rhizobial story has an additional delightful twist. In 2014 three teenage Irish girls won a globally based science research competition at the GoogleScience Fair, based in San Francisco. Sophie Healy-Thow, Emer Hickey and Ciara Judge, from Kinsale

Community School, in County Cork, were named grand prize winners in the 15- to 16-year-old category for a project aimed at using 'natural bacteria' to increase crop output. Inspired by hearing about famine in the Horn of Africa in 2011, they set out to find some way of helping to boost food production in Third World countries. Using naturally occurring rhizobia known to fix nitrogen, they converted Ciara's home and garden into a temporary laboratory, testing thousands of different plant seeds, and looking to see what happened when they added the nitrogen-fixing rhizobia to the soil. Through systematic measurement and observation, they discovered that by simply adding the rhizobia to the soil they could speed up the germination processes of high-value crops such as barley and oats, boosting output by up to 50 per cent.

Their prize included $50,000 towards further funding their project, which now included exhibiting their research, in association with National Geographic, to the Galapagos Islands. It also included the offer of a future trip to outer space as part of Richard Branson's astronaut project.

19

The Ecology of the Oceans

On a pleasant September day, back in 1994, I arrived at the Rockefeller University, New York, to interview its distinguished president. Back in 1958 he had been one of the pioneers of the study of DNA, sharing the Nobel Prize in Physiology or Medicine with Edward Tatum and George Beadle for advancing our understanding of bacterial genetics. In essence what Lederberg and Tatum discovered was the fact that bacteria had a form of sex life.

Up to then microbiologists had assumed that bacteria exclusively passed down their genetic information through the process of budding reproduction. But if this were so, future generations of bacteria would all be genetic clones of the original strain. What Lederberg and Tatum discovered was that genetic material can be passed from one bacterium to another through a process known as 'conjugation'. This process might be seen as the bacterial equivalent of sexual reproduction in plants and animals, since it involves direct physical contact between the donor and recipient bacteria, during which a packet of genetic material is transferred from the donor to the recipient through a bridging structure called a 'pilum'. This is just one of three different mechanisms through which bacteria acquire new genetic information. A second mechanism, termed 'transformation', involves the bacterium absorbing new genetic information directly through the bacterial cell wall. This could only take place if other bacteria have been destroyed, for example through viral lysis, with their genetic material being released into the ambient medium. Another mechanism, known as 'transduction', involves the introduction of new genetic information in the form of an invading phage virus. These intimate interactions between bacteriophage viruses and their microbial prey is currently opening up a new era of understanding of what is going on within the Earth's biosphere.

At the time of my interview with Professor Lederberg, I was investigating the theme of emerging plague viruses. To my surprise, I came to appreciate that viruses were not the simple genetic parasites that most doctors assumed. I came to the Rockefeller to ask one of the most knowledgeable experts in the world a relevant question. Could it be that viruses were not merely parasites but true symbionts in their interactive relationship with their hosts? It turned out to be a fascinating interview that Professor Lederberg graciously allowed to take up most of the afternoon. In reply to my question, he confirmed that, at least when it came to phage viruses and their bacterial hosts, viruses sometimes behaved as one would expect of symbionts. He also told me that he had no knowledge whether or not this applied to the rest of nature, but he encouraged me to check it out. I took his advice. The result of my exploration was positive, with evidence for the symbiotic behaviour of viruses being confirmed in many fields of virology.

Some of the most important, even astonishing, confirmations

came from the very lines of work initiated by Lederberg and those earlier workers in the fields of microbiology, and in particular from links between the expanding sciences of virology and genetics that Lederberg and his colleagues helped to pioneer. Today there is abundant evidence that viruses are the ultimate symbionts. They are obliged by their very nature to enter into a wide spectrum of symbiotic relationships, including parasitism, commensalism and mutualism, with their specific hosts. Sometimes, as we saw with myxomatosis in the Australian rabbits, the relationship begins with an aggressive parasitism, only to metamorphose into an eventual mutualism. Back in 1974 Joshua Lederberg was already convinced that bacteriophage viruses were capable of entering into such symbiotic relationships with their bacterial hosts.

One key ecology in which viral symbiosis is now proving extremely important is the oceans, which occupy some 77 per cent of the Earth's surface area. But their actual contribution to the world's living space is much greater than that since the oceanic ecology is three-dimensional. The dominant organisms in the oceanic ecology are not killer whales, shoals of tuna or hunting sharks, not even the myriad colourful denizens of the coral reef: they are much simpler organisms, for the most part the prokaryotes described earlier, which consist of bacteria and Archaea, together with the microscopic, unicellular eukaryotes known as protists. The vast populations of these microscopic life forms are the true foundation of the marine food web. Only in the last decade or so have microbiologists belatedly come to realise that the viruses that 'infect' these same microscopic life forms also constitute a major component of the biosphere. This new understanding, which one authority has labelled 'the great virus comeback', has emerged through the expanding fields of evolutionary biology, genetics, genomics, metagenomics and population dynamics, opening our vision to the fact that viruses are both essential interactive agents within the tree of life and key to understanding the complex dynamics of major ecologies.

To explain why viruses are vital to the ecology of the oceans, we need to look more closely at the nature of the symbioses involving viruses and their host bacteria. These symbioses involve two very different patterns of interaction, known as lysogenic and lytic cycles. Both these interactions are intrinsically linked to viral reproduction, but while one is described as 'temperate' the other is exceedingly aggressive. When an invading phage virus enters into a lysogenic, or 'latent', cycle with the bacterial host, the virus integrates its genome into the host genome, or else it sits outside the genome in the form of a circular 'replicon' within the bacterial cytoplasm. While behaving temperately, the virus merely holds its position as a 'prophage' without attempting to complete the lytic viral cycle of replication. Meanwhile, the viral genome is replicated whenever the bacterium reproduces, through its budding pattern of reproduction, with the prophage being passed on to the daughter bacteria. But all the while the virus retains its lytic potential and, over time, certain stimuli can provoke a change to the more aggressive pattern of behaviour.

In the lytic cycle the virus inhabits the bacterial cell as a separate genetic entity, but in this case the virus replicates within the bacterial body independently of bacterial reproduction, through hijacking the bacterial genetic apparatus for its own selfish purposes. Viruses that behave in this way are known as 'virulent' phages. They create a swarm of daughter viruses within the host bacterium, with the ultimate death and rupture of the bacterial cell releasing the daughter viruses into the surrounding medium, where they are poised to infect large numbers of ambient host bacteria. This process of virulence, swarm type replication and subsequent release with rupture of the bacterial cell, is known as 'lysis' and the cycle is known as 'the lytic cycle'.

Sometimes the situation can become even more complicated. For example, bacteria sometimes manage to shrug off their unwanted phage passengers during reproduction. But phage viruses have evolved a cunning strategy to counter any such escape. A phage species known as P1, which is commonly found in nature, adopts a defensive strategy called 'an addiction module', which relies on the presence of two antagonistic genes. One of these is expressed within the host bacterium as a stable toxin that is potentially lethal to the bacterium: meanwhile the second gene is expressed as a short-lived antitoxin that opposes the lethality of the toxin. If the bacterium manages to shed the phage virus during reproduction, the life-saving action of the short-lived antitoxin is soon lost, exposing the daughter bacterium to the lethal effects of the persisting toxin. This guarantees that only bacteria that retain the highly manipulative virus are allowed to live and reproduce.

Today we know that the oceans not only teem with microscopic life forms, consisting largely of bacteria, Archaea and protists, they also teem with the bacteriophage viruses that prey upon them. Over the last two decades we have come to realise that these phage viruses, with their intimate interactions with their host prokaryotes, are key players in the ecologies of the oceans. Writing a review of this emerging field, in 2005, Curtis A. Suttle declared: 'Viruses exist wherever life is found. They are major causes of mortality, a driver of global geochemical cycles and a reservoir of the greatest genetic diversity on Earth.' A litre of surface seawater typically contains somewhere between ten and a hundred billion viruses. But this is no cause for alarm for you and me; the vast majority of these viruses are bacteriophages, often unidentified and unstudied to date, which have no interest in us. These myriad bacteria, Archaea and protists play a key role in the biosphere, fixing key elements such as carbon, nitrogen and phosphorus into organic chemical compounds. Phage viruses exclusively infect these microbes, killing them through gargantuan cycles of lysis, and thus recycling their stores of nutritional elements to become the foundation of the oceanic food chains. These colossal phage lytic cycles are key to keeping the balance of nature in the oceans, preventing the build-up of toxic microbial blooms and recycling carbon and other nutrients back into the nutritional webs of other microorganisms.

To our human senses, these never-ending natural cycles of life and death might seem brutal, perhaps even counter-intuitive. But as Charles Darwin realised, some century and a half ago, nature is decidedly not benign. We only need to reflect on everyday life to realise that we are part of the same recycling from birth to death, which is the norm for all sexually reproductive life on Earth. Today we know that phage viruses are among the commonest and most diverse of biological entities in the biosphere. It has been estimated that there are more than 10^{31} such viruses on planet Earth. That is some ten million trillion trillion, which adds up to more than all the other organisms put together, including bacteria, by an order of ten-fold to a hundred-fold. These viruses are voracious predators of microbes, yet they also play an important role in the evolution of the same microbes – and clearly an enormously important role in the ecology of the oceans.

The more we look for viruses in all manner of ecologies, including the anoxic deeps of the human intestine, the more we find them, often hitherto unknown varieties of viruses whose function in their respective ecology is largely unknown. At the same time, we are also just beginning to query the role viruses might have played in the earliest stages of the evolution of life, much as they continue to play a key role in the interdependency of life today.

Viruses are quintessentially different in their life cycles and genetic make-up from the three great domains of cellular life. Yet at the same time viruses interact at every level within those three domains because of their symbiotic dependency on cellular life for their very existence. This combination of that quintessential difference, and the obligatory nature of their interactions with their cellular hosts, is the key to understanding the symbiotic nature of viruses. The fact that viruses are obliged to hijack the genetic and metabolic pathways of their hosts, coupled with their remarkable dexterity in manipulating these same pathways, gives viruses the potential for changing host evolution – through a pattern of evolution called 'genetic symbiosis'. Through genetic symbioses with host-genome-changing viruses, cellular life has benefited from a potential for novel evolution that would not have evolved without the benefit of the viral contributions.

Only recently have we come to realise that the 'virosphere', which comprises the junctional zones where viruses interact with their myriad hosts, spans all environments where life is to be found. In every ecology where we have systematically searched for the presence of viruses, we have found them to be the most abundant biological entities. We are only coming to realise that the genetic diversity of viruses – the range of different viral genes and genetic sequences – is also commensurately enormous and may substantially exceed the genetic diversity of all the cellular organisms combined. This was confirmed by a recent metagenomic analysis of the marine virosphere of four different oceanic regions, which revealed that most of the viral sequences found in these ecologies are unlike any of the known sequences in current genetic databases. So high was the genetic diversity found to be, it suggested the presence of several hundred thousand presently unknown viral species. Among the newly discovered symbiotic interactions involving viruses, some were contributing to the evolution of the cyanobacteria, which are important to photosynthesis, which in turn is vital to the energy cycles of the oceans as well as to the oxygen we breathe.

Until recently most of the studies of viruses in the oceans were conducted on samples from surface waters. These confirmed that viruses were the most abundant biological organisms in every ocean tested. But little was known about the presence, and role, of viruses in deeper oceanic ecosystems. In 2008, an international group of marine scientists reported a pioneering study of the viral impact on benthic deep-sea ecosystems, confirming that viruses are abundant at all depths, from the oceanic surface waters to the level of abysmal sediments. Further experiments showed that more than 99 per cent of the viral infection cycles at every level were lytic in nature, including the benthic boundary level and even the deep-sea sediments. Everywhere they were observing those gargantuan killing cycles in action.

The study of viruses in oceanic ecologies is still early, but already it would appear that viruses are playing an important, indeed essential, role in the oceanic geochemical cycles. This begs an important if altogether predictable question: is this ecological role of viruses confined to the oceans?

20

The Virosphere

In 2006, I helped to organise a half-day session on the theme of viruses as symbionts at the International Symbiosis Society's world meeting in Vienna. It was the first time this theme had been presented in the history of the society. The eminent evolutionary virologist, Professor Luis P. Villarreal, chaired the session and we were joined by American botanical virologist, Professor Marilyn J. Roossinck, then working at the Samuel Roberts Noble Foundation, in Ardmore, Oklahoma. The large theatre full of non-virologists was somewhat astonished to hear about the expanding field of viral symbiosis, but by and large they welcomed our contribution to their discipline while worrying somewhat about what appeared to be an additional layer of complexity that was likely to complicate many symbiotic interactions.

A year earlier Marilyn Roossinck had emailed me to say that she was interested in the views I had presented in a book, *Darwin's Blind Spot*, in which I had compared and contrasted the neo-Darwinian approach and symbiological approaches to evolution. Attached to her email was a review she had written for the journal *Nature Reviews*. It was a very interesting paper in which she compared and contrasted the same two theoretical approaches with respect to the evolution of novel viruses. She concluded that both mechanisms were capable of giving rise to new species of viruses, but evolution through symbiogenesis was the most likely model for many of the evolutionary events that had resulted in rapid changes or the formation of new species. Her review was published in December 2005.

In 2007, Professor Roossinck and colleagues at the Samuel Roberts Noble Foundation conducted a pioneering field experiment in Yellowstone National Park, in Wyoming. Botanists had already shown that symbioses between plants, such as tropical panic grass and fungi that invaded the plant tissues, helped some grasses to survive in this arid, high-temperature ecology. Fungi, like all other life forms, are susceptible to infection by viruses. The researchers now focused on a single species of tropical panic grass found in the geothermal soils of Yellowstone, called *Dicanthelium lanuginosum*, and its symbiotic partner fungus, *Curvularia protuberata*. Previous experiments had shown that neither plant nor fungus alone was capable of surviving soil temperatures above 38°C, but they thrived when present together as a symbiotic association. Nobody had previously considered the possibility of a viral presence. But now when they screened the plant–fungus association for the genetic sequences associated with mycoviruses, they confirmed the presence of an unknown virus. This raised the possibility that the association might have a third contributing member – a symbiotic virus that was helping the plant–fungus association survive in the arid ecology.

When they 'cured' the fungus of its virus, they discovered that the fungus was no longer capable of conferring heat tolerance on the grass. The symbiotic nature of the virus was further confirmed when its reintroduction restored the heat tolerance. They published this iconoclastic research under the title: 'A Virus in a Fungus in a Plant: Three-Way Symbiosis Required for Thermal Tolerance'. A year later, the same researchers extended their work to discover more instances of drought tolerance in various plant species inoculated with four different viruses. In their summary, they no longer referred to viruses as 'obligate genetic parasites' but as 'obligate intracellular symbionts'.

More recently, Professor Roossinck and her colleagues have moved on to make use of metagenomic searches to explore nature for viruses in plants. In 2011, in a paper entitled, 'The big unknown: plant virus biodiversity', they predicted that comprehensive screening studies would reveal a picture of plant virus diversity well beyond our current understanding. In that same year, a multi-centric group of virologists published a startling overview of a situation they referred to as 'the prokaryotic virosphere'. They began by affirming that, over the previous few years, the prevailing perspective of the viruses of prokaryotes had been transformed from mere laboratory interest to become a major consideration in our understanding of major ecosystems and indeed the planetary biosphere. It now looked increasingly likely that widespread interaction between Eubacteria and Archaea and their viruses has played a vital ecological role for billions of years, possibly for as long as cellular life has existed on Earth.

What could have persuaded such knowledgeable scientists to arrive at such ground-breaking conclusions?

To begin with, microbiologists now realised that the measurable population numbers of these prokaryotic viruses over the whole of the planet were astoundingly vast. And even the prevailing measures in the oceans were likely to have underestimated the total numbers of viruses on the planet, since they only screened for tailed bacteriophages, which are the easiest to measure. Such screening excluded numerous other major groups of bacterial and archaeal viruses, and it ignored the subtler presence of viruses embedded in the genomes of bacteria and Archaea as 'proviruses'. Other groups of researchers, including microbiologists and biotechnologists, had begun to extend the same types of screening for the presence of viruses in soil.

Up to then, most plant virologists had been preoccupied with the role of viruses in plant diseases, and their implications for crop production. But by 2005, plant and soil scientists from the Universities of Delaware and Tennessee set out to examine the abundance and diversity of viruses in six different ecologies of Delaware soils, where they duly confirmed a similar abundance to what had been found in the oceans, with billions of viruses per gram dry weight of soil. As with the oceanic studies, they found that the viral communities in soil were dominated by bacterio-phages. They also found a higher abundance of viruses in forest soils when compared to agricultural soils and, predictably, they confirmed that virus numbers correlated with bacterial abundance and with moisture and organic matter content rather than with any particular soil texture. Even more astonishing was the discovery, by this same group, that viruses were also abundant, if reduced to hundreds of millions per gram of dry soil, in the rela-tively lifeless Antarctic valleys.

This surely rings a bell with the findings of the Megavirus hunters.

By now microbiologists were waking up to the importance of these hitherto unknown virus–microbial symbioses in multiple ecologies. In an overview, published a year after the Antarctic study, a new paper took stock of the ecological importance of virus population recycling in the coastal environments, which contributed to something between 20 to 100 per cent of bacterial stock turnover. This appeared to imply the same ecological impli-cations as the colossal cyclical lysis of bacterial populations further out in the oceans, enabling transfer of key elements such as carbon and iron and other micronutrients from the bacterial biomass to smaller prokaryotic life forms within the same environments. Where, up to recently, little had been known of the ecological importance of viruses in soil, now the findings of Williamson and others in varieties of different soils and environments suggested that, despite the predominance of aquatic environments on planet Earth, the microbial abundance and diversity within soil environments might well exceed that of the aquatic realm.

In a 2017 review with the evocative title, 'Viruses in Soil Ecosystems: An Unknown Quantity Within an Unexplored Territory', Williamson and his colleagues emphasised that soil virus diversity was still very much underestimated and the impact of viruses on soil ecosystems was poorly understood. Two years earlier, Marilyn Roossinck had come to the same conclusions: there was a pressing need for advances in plant virus metagenomic studies.

Today new metagenomics studies of virus populations in various aqueous ecologies are producing interesting results, with tested sites including man-made and Antarctic lakes, Chesapeake Bay, numerous aquaculture systems, the hot springs of Yellowstone National Park, and the hydrothermal vents on the bottom of the oceans. Metagenomic explorations are also expanding into different soil ecologies, including the Kogelberg Nature Reserve in the Cape Floral Kingdom of South Africa, the rain forests of Peru, the Californian desert, the Kansas prairie and rice paddy soils of Japan and Korea. Indeed, some researchers are exploring what might appear to be the limited if exotic ecology of the human virome.

Our bodies embrace mobile ecologies that we live with from day to day and carry about with us on our travels. For many of us these are personal ecologies that we prefer not to think about, yet these same ecologies are important to our normal health and well-being. One such personal ecology, perhaps the most mysterious of them all, is occupied by those lurker viruses that hide somewhere within our physical being, such as herpes simplex, herpes zoster, glandular fever and cytomegalovirus: all those viruses that seem to stick around with us for life. Other more obvious ecologies are those microbiomes mentioned in the early chapters of this book, such as the skin in general, and especially the more moist areas, such as the armpits and groin, the nasal cavity and the mouth, extending also, in females, to the vagina and genital tract and, in both sexes, to our largest and most obvious microbial ecology of all: the large bowel, or colon, where vast numbers of microbes teem and contribute to human health.

Some 100 trillion microbial cells are found in the colon. Given what we are learning about the various ecologies of the planet, we should hardly be surprised to learn that the colonic microbes have also attracted the concomitant presence of viruses in teeming abundance. The simplest way to look for the colonic viruses is through the examination of the evacuated stools. Stools from healthy individuals contain up to one hundred billion microbial cells in a single gram. These are mainly bacteria but they also include Archaea and protists. Studies of the viral component of the colonic microbiome are at an early stage, but they already confirm a dynamic symbiotic inter-relationship.

Allowing for the fact that study of the human virome is at an early stage, we already know that the healthy colon is inhabited by enormous numbers of phage viruses, with genetic sequences for the most part unknown in any of the established genetic databases. The fact that these viruses are referred to as 'viral dark matter' does not portend anything sinister. These viruses are unknown because, until recently, nobody bothered to study them. Each of us lives in a healthy harmony with these massive numbers of bacteria, and their symbiotic viruses, from soon after birth. Currently, like the viruses in all those other viromes on Earth, these are being subjected to metagenomic analysis. They are also being studied to see if specific populations can be related to various aspects of health and disease.

One such study aims to correlate the colonic virome with variations in diet. The initial results suggest that each of us harbours a uniquely personal and fairly stable virome over lengthy periods, meanwhile the viromes of different individuals on the same diet do tend to converge in detectable ways. This suggests that diet does influence our virome composition.

Other studies have taken advantage of the fashion in some quarters for faecal microbial transplantation. This has led to a series of experiments that are currently looking at the impact of faecal microbial transplantation on the colonic virome. For example, some success has been reported for this procedure in treating relapsing infection of the gut with the food-poisoning bacterium, *Clostridium difficile*. Faecal microbial transplantation has also been tried as a treatment for ulcerative colitis in children, where a limited transient benefit was reported. It would appear that, like all of life on Earth, we live within, and are inhabited by, ecologies that teem with viruses. Put simply, life on Earth has evolved in a never-ending, profoundly symbiotic interaction with co-evolving viruses. We humans, like all of cellular life, inhabit and are inhabited by, Earth's virosphere. This extraordinary finding, with its manifold intimate symbiotic interactions, in the oceans, in the soil, inside our very body cavities, is likely to have contributed to the history of life on Earth, to biodiversity as it currently exists, and to its continuing evolution.

One particular family of viruses has come up very close and personal indeed to our human evolution.

21

The Origins of the Placental Mammals

The twentieth century saw the emergence of the most dangerous new virus to afflict humanity since smallpox. We know it as the human immunodeficiency virus, or HIV-1, and we are all too familiar with the lethal pandemic caused by its emergence as acquired immunodeficiency syndrome, or AIDS. We also know where the virus came from. It is closely related to a virus of chimpanzees, called simian immunodeficiency virus, or SIV. Both HIV-1 and its chimpanzee cousin are retroviruses. In other words, they are RNA-based viruses that contain their own viral enzyme, reverse transcriptase, which enables the virus to convert its RNA-based genome to the equivalent DNA template during the process of infecting its target cell within the host body. That target cell, as we have seen so very often with viruses, is a cell intrinsically involved in the immune response to the viral invasion, on this occasion a T-lymphocyte. HIV-1 attaches itself to a specific receptor on the surface membrane of the T-lymphocytes, known as the CD4 receptor, which allows the viral envelope to fuse with the cell membrane, facilitating entry of the viral genome into the cell interior. Here, making use of its reverse transcriptase enzyme, HIV-1 converts its genome into DNA, which is inserted into the chromosomes of the lymphocyte to become a template for daughter virus production. This viral template within the host chromosomes is called a 'provirus'. The provirus instructs the cell's genetic machinery to manufacture new daughter viruses, which are released into the surrounding tissues, and ultimately the blood-stream, where they repeat the replication process over and over with other host T-cells, sometimes widening the host cell target spectrum to infect other lymphocytes, macrophages, dendritic cells and even brain cells, all of which are presumed to possess CD4 receptors on the cell membranes.

The full-blown syndrome of untreated AIDS causes massive damage to the human immune defences. This in turn results in life-threatening secondary infections with opportunistic organisms including *Cytomegalovirus*, *Toxoplasma*, *Candida*, herpes simplex and a number of other microbial agents that would not normally be capable of causing such overwhelming infection in a person with an intact immune system. Another complication is the cancer known as Kaposi's sarcoma, which affects the skin and the internal organs. The cause of AIDS remained a puzzle until the HIV-1 was discovered by Luc Montagnier and Françoise Barré-Sinoussi at the Pasteur Institute in Paris in 1983.

Viruses, such as HIV-1, do not come out of thin air. Emerging viruses come from pre-existing sources, often through human intrusion into long-standing virus–host symbiotic cycles in nature. We now believe that HIV-1 was acquired through a cross-species jump from chimpanzees that had long been infected with a simian immunodeficiency virus, or SIV. A second, less virulent immuno-deficiency virus, HIV-2, was probably contracted from a similar species jump from sooty mangabey monkeys, which are hosts to a closely related strain of SIV. By now we are hardly surprised to discover that, in both these animal hosts, the viruses rarely if ever cause symptoms of disease.

How might such rain-forest viruses have jumped from chimps and monkeys to infect humans?

The most likely explanation is through hunting. Local African populations have a tradition of hunting apes and monkeys for bushmeat, which would have brought humans into contact with the blood of apes and monkeys that was infected with immuno-deficiency viruses. The viruses are likely to have gained entry to the tissues of the hunters through cuts or abrasions on their skin. This pattern of species crossover was confirmed in 1999 when researchers found a strain of SIV in chimpanzees, known as SIVpz, which was almost identical in its genetic sequences to HIV-1 in humans. Through back-tracking the ancestry of HIV-1 in stored human blood, geneticists confirmed that the first human case of HIV-1 infection was likely to have been a man living in Kinshasa in what is now the Democratic Republic of the Congo. They concluded that the species crossing to humans took place some-time in the early 1920s. From the DRC, AIDS made its way to Haiti during the 1960s, when Haitians who had been working in the DRC returned home to the Caribbean island.

Over subsequent decades AIDS spread to America, Europe and ultimately, worldwide. By this stage, HIV-1 had evolved into a number of different 'strains' or 'subtypes', which spread in a variety of different patterns of contagion in its human host: for example, in gay men or through contaminated needles in illicit drug-taking, through heterosexual sex, or from mothers to babies. By the year 2016 the 'M' strain – M for major – had infected some 75 million people. Roughly one million died from AIDS that same year, though this was a considerable improvement on the figures for 1997, when the global mortality peaked at 3.3 million.

Today, through high-quality surveillance, advice and measures on preventing spread to sexual partners and family members, combined with effective multi-drug therapy, AIDS is no longer the death sentence of former years. Some call this a 'functional' cure, but it is not yet a complete cure, in the sense of the eradication of the virus from an infected patient. In the year 2016, 36.7 million people were still living with the presence of HIV-1, comprising roughly equal numbers of men and women. How long before medical science discovers that highly desirable sterilising cure? We just don't know at present. Considering its tiny physical size, and its equally tiny genome, HIV-1 has shown an extraordinary capacity to resist therapeutic annihilation. How, one is inclined to wonder, has this minuscule entity managed to confound all that the modern world of microbiological and therapeutic research and knowledge could throw at it for what is now three decades of antiviral blitzkrieg?

We might consider this viral persistence somewhat perplexing. Surely the first arrival of HIV-1 into the human blood and tissues should be picked up by our powerful immunological surveillance systems? We doctors have seen how, in the case of many other pathogenic organisms, this system spots and then mops up viruses using antibodies and the various cellular elements of the immune system, much as we saw with the common cold, or the norovirus, or with the phage viruses I injected into rabbits and humans many years ago. Is there something different about HIV-1 and its battles with our immunity?

Retroviruses are very ancient viruses, older by far than the mammals, older even than the vertebrates found in the earliest fossils. They have had a great deal of time in which to hone their abilities to outwit the immune defences of their hosts. One such tactic employed by HIV-1 is to hide its presence inside a kind of 'invisibility cloak' made up of our own human molecules, so it is not recognised as 'alien' by the human immunological defences. Another key to the success of HIV-1 in its pandemic behaviour is the virus's mutational capacity. Retroviruses, like all RNA viruses, have a ferocious capacity for mutation. By 1985, some five or six years after the AIDS pandemic was recognised, viruses infecting individual AIDS patients were already showing envelope gene sequence variations that were 12 per cent different from the initial sequence. Six years later, AIDS patients in Florida were now showing a staggering 19 per cent variation. HIV-1 mutates so lightning fast inside every infected patient that the dominant strain of virus within a single individual actually changes during the course of his or her infection. In a sense, every sufferer evolves his or her own strain of virus, and, within each sufferer, the strain is not a single viral genome but swarms of related versions of the virus, with each individual swarm actively supporting its own swarm relatives, meanwhile, the swarms are furiously mutating and competing with one another.

Moreover, HIV-1 and HIV-2 are not the first retroviruses to afflict humanity. Study of the human genome reveals that retrovirus epidemics have struck our human ancestors again and again. Even further back, they have infected our pre-human primate ancestors, and every ancestral stage going back to the very origins of the vertebrate animals, with evolutionary implications that are frankly extraordinary. To understand what this signifies we need to grasp how retroviruses replicate themselves inside the host target cell.

We might recall that retroviruses use the viral reverse transcriptase enzyme to convert their RNA genome to its equivalent DNA sequences, which they then insert into the chromosomes of the target cell, using the virus's own 'integrase' enzyme. It is these spliced retroviral genomic inserts inside the genome of the host target cell that serve as 'proviruses' coding for daughter viruses. Sometimes, during a retroviral epidemic in a new host species, the same viruses use exactly the same techniques to insert their viral genomes into the host germ line cells – the ova and sperm that will become the future offspring. When this happens the inserted viral genomes will be inherited, like any other genetic sequences within the germ line genome, into future generations of the species. Retroviral genomes are powerfully interactive from the genetic point of view.

One might shake one's head and wonder what, if any, effect this arrival of a viral genome into the host chromosomes could have on the future evolution of the infected species. But we need to consider that these viruses are symbionts of the same host. They have evolved the capacity to manipulate the host physiology and genetic machinery. The presence within the germ line of viral genes, and regulatory regions, known as LTRs, will have numerous potentials for changing the future evolution of the host species. This is a cardinal example of evolution through 'genetic symbiosis'. We are currently observing retroviruses by inserting their genomes in exactly this way in a prevailing retrovirus epidemic of koalas in Australia.

A little more than a century ago a retrovirus jumped species, very likely from a rodent, to give rise to an epidemic infection of the koalas in eastern Australia. Like AIDS in humans, the disease was spread through sexual intercourse. We can track its movements judging from the observation that virtually all of the koalas in northeastern Australia are now infected; about two-thirds of those half-way down the coast are similarly infected; and a third of the koalas in the south. Koalas that were introduced into an island off the eastern coast, known as Kangaroo Island, a little more than a century ago, remain unaffected. This suggests that the epidemic began in the northeast and has been making its way south over more than a century. It seems inevitable that all of the Australian koalas, other than a small number cut off by geographic isolation on islands, will end up infected, illustrating the remarkable efficacy of sexual transmission of a retrovirus. Behaving like HIV-1 in humans, the koala retrovirus has killed millions of animals, largely through leukaemias and lymphomas. Meanwhile, the virus is 'endogenising' into the koala germ line, with some animals already accumulating up to 100 'provirus loci' scattered throughout their chromosomes. Study of this extraordinary plague virus behaviour is illuminating how retroviruses have played a key role in the evolution of animal genomes, including our own.

If we examine the retroviral component of mammalian genomes, we discover vast numbers of provirus inserts scattered throughout the chromosomes. These viral inserts are known as 'endogenous retroviruses', or ERVs. Endogenous retroviruses are found in the genomes of every vertebrate. They preceded the arrival of the first land vertebrates, being discovered in amphibians and fish, shark and frog. Retroviruses of even more ancient origins have been found in the photosynthetic sea slug, *Elysia chlorotica*, where the virus floods the tissues of the slug populations close to the end of their annual cycle. When geneticists examined the *Elysia* retrovirus genome, they found sequence similarities to retrotransposon sequences found in a sea slug, called the Californian sea hare, and the purple sea urchin, both of which inhabit the Pacific coasts of the United States. These findings suggest that retroviruses are very ancient indeed and they are likely to have played important roles in the evolution of the entire animal kingdom. We find extraordinary confirmation of this role when we examine the contribution of endogenous retroviruses to our human evolution.

Our chromosomes contain an astonishing 203,000 retroviral provirus inserts, that are the result of at least two hundred or so retroviral epidemics in the lifetimes of our human and pre-human ancestors. Over time this retroviral inheritance has radically changed our human evolution. The key to understanding this pattern of evolution is to grasp how a genetic symbiotic pattern of evolution actually works, and in particular to grasp the concept of holobiontic genetic evolution.

When a viral genome becomes embedded in a host germ line, the two genomes unite to form a new 'holobiontic genome' – a genome that contains more than one evolutionary lineage. In consequence, the 'holobiontic genome' will have new potentials for evolution that depend on the interactions between the former host genome and the highly manipulative viral genomes. Darwinian natural selection will no longer operate exclusively at a selfish level on either virus or human component but also at the level of host–viral holobiontic genome. It will select for genomic changes that enhance the potential for survival of the holobiont and will select against changes that reduce the potential for survival of the holobiont, regardless of whether these changes apply to genes and regulatory elements of former host or former virus.

In the human genome, retroviral inserts are referred to as 'human endogenous retroviruses', or HERVs. Human endogenous retroviruses comprise between 30 to 50 families, depending on definition, and these families are further subdivided into more than 200 distinct groups and subgroups. Each of these is thought to represent an independent invasive viral lineage, confirming that our primate ancestors have been the victims of a large series of retroviral epidemics. Although most of these epidemics took place more than 10 million years ago, a significant number happened after the separation of the human lineage from that of chimpanzees, perhaps seven million or so years ago. At least ten of these endogenous retrovirus colonisations, known as HERV-Ks, are unique to humans.

Over the lengthy period of holobiontic evolution, there will be opportunities for evolutionary interaction between the viral inserts and the host genome for evolutionary advantage. One way in which the proviral inserts might change the human genome is through offering new genetic control capability arising from the vast numbers of insertions of viral regulatory regions, known as LTRs, especially when they lie somewhere close to human genes. Is there hard evidence from the examination of the human genome today for selection working in this way at holobiontic level?

The answer, overwhelmingly, is yes!

Without going into genetic details, large numbers of formerly viral genetic regulatory regions are now actively controlling the transcription of human genes into their respective proteins. A systematic screening of different classes of regulatory regions in the human genome discovered key viral genetic sequences that were affecting the function of some 533 human genes. For example, the regulatory region of an endogenous virus known as ERV-9 has replaced the former host controls of the β-globin gene cluster – a group of five genes that code for the beta globin of the haemoglobin in our blood.

In the year 2000, two separate research groups discovered that the envelope gene of a human endogenous provirus locus, known as ERVWE1, is essential for the structural formation of the human placenta. The ERVWE1 locus is inserted into human chromosome 7. Its viral envelope gene, or *env*, which would normally have coded for the protein component of the viral envelope, now codes for a protein called syncytin-1. Syncytin-1 is strongly expressed in the human placental interface cells, known as trophoblasts, which fuse together to form a syncytium – a single confluent membrane with no junctional gaps between adjacent cell walls. In effect, syncytin has altered the fate of trophoblasts, converting them into syncytiotrophoblasts. This syncytium, which is an extremely thin membrane, forms the interface between the maternal and foetal circulation, invading deep into the lining of the womb during pregnancy. Because there are no gaps between cells this ensures that all the nutrients from mother and waste from the foetus pass through cytoplasm and thus are biologically filtered. Half the foetus antigens come from the father and so would be seen as 'foreign' by the maternal immune system. The confluent syncytial cell layer separating the maternal from the foetal circulations in the placenta also helps protect the foetus from attack by the maternal immune system.

This placental–endometrial interface is one of the most refined of all the mammals. We share it with the great apes, including gorillas, orang-utans and chimpanzees, hence the viral locus is an example of an 'ERV', or endogenous retrovirus, rather than the humanly restricted 'HERV', or human endogenous retrovirus. A second endogenous retroviral protein, dubbed syncytin-2, is expressed by a provirus locus, HERV-FRD, on chromosome 6. Syncytin-2 is expressed on the foetal side of the placental interface layer, where it expresses a powerful immunosuppressant function, helping to protect the foetus from maternal immunological attack. Today we know of at least 12 different endogenous retroviral loci that play significant roles in human reproduction, with at least five involved in placentation, though the precise roles of some of the others remain to be determined. Indeed, we are only beginning to appreciate the contribution of retroviruses to human reproduction, embryological development, immunology and cellular physiology.

Soon after their discovery in human placentation, the key functions of syncytins 1 and 2 type proteins were discovered to play similar roles in other mammals. For example, mice were discovered to possess two very similar genes, syncytin-A and syncytin-B, which function during placentation in much the same way. To test this, scientists bred a generation of mice with defective expression of syncytin-A and syncytin-B. They found that their embryonic placentas now displayed major defects in cell-to-cell fusion and invariably resulted in the death of embryos. This confirmed that syncytins, coded by viral loci in mammalian genomes, are essential for normal placental structure and function.

The study of the syncytins, and the various other contributions of the huge assortment of endogenous retroviruses within the human genome, is only just beginning. But it is already clear that they have made a profound contribution to our human evolution. There is growing evidence for the fact that so-called viral envelope genes are also involved in many different human cells, tissues and organs, so much so that scientists are now beginning to work on what is called the 'HERV Transcriptome'. Indeed, the huge viral component of the human genome has potential for both good and harm. We are at an early stage in figuring out that endogenous retroviruses play a vital role in embryogenesis. However, sometimes aberrations of the syncytins and other endogenous retroviral genes may also be involved in human placental abnormalities, some aspects of Down's syndrome, diseases of pregnancy such as pre-eclampsia, intrauterine growth retardation and a related form of cancer known as choriocarcinoma. More widely, endogenous retroviruses are believed to play some role, for good or for bad, in some of the autoimmune disorders and many different forms of cancer.

Indeed, symbiogenetic retroviruses would appear to confront us with a very important dilemma: just how critical a role have retroviruses played in the origins and evolution of the placental mammals?

Following the discovery of the two types of human syncytins, some French researchers, led by Thierry Heidmann, set out to answer this question. They screened a large number of different mammalian groups for the presence and function of the two key retroviral syncytin type genes. The results were astonishing, confirming that a variety of syncytin *env* gene variants were indeed playing similar placental roles in every group they studied. These included the great apes, with syncytin-1 and syncytin-2, and the rodents, with their syncytin-A and syncytin-B. To these they now added the lagomorphs, which include rabbits; carnivores; horses; bats; ruminants; cetaceans, which include whales, porpoises and dolphins; the *Suina*, which includes pigs; *Insectivora*, which includes hedgehogs and shrews; *Afrotheria*, which include elephants, aardvarks and sea cows; and *Xenarthra*, which include anteaters, sloths and armadillos. In every group tested, they discovered variants of the two key retroviral syncytins.

The French scientists did not stop there. They turned their attentions to the marsupials, a group of animals closely related to mammals but which do not employ full-term placentation as part of their reproduction. Some marsupials, such as the South American opossum, are known to employ a very short-lived stage of placentation during the time when the foetus is coming down into the pouch. When the French scientists examined the opossum's genome for the presence of those same key syncytins, they discovered a new syncytin-1 type of gene, which they labelled syncytin-Opo1. Probing further, they found a second retroviral envelope gene that had been selectively conserved for more than 80 million years among all marsupials, including the South American opossum and the Australian tammar wallaby. This second envelope gene had immunosuppressive properties. In other words, it appeared to function much as synctytin-2 does in great apes.

Up to this point there was uncertainty whether the viruses had played a key role in the very origins of placentation, or whether they had arrived after the evolution of a more primitive placenta and helped to make it more efficient. The discovery of the two key retroviruses in marsupials with a transient placentation answered this dilemma. In the researchers' summary: 'The capture of a founding syncytin by an oviparous ancestor was pivotal for the emergence of placentation more than 150 million years ago.'

The implications are altogether clear. In plain English: no retroviruses, no placental mammals.

22

Viruses in the Origins of Life

We inhabit a world of mystery and wonder, even if, distracted by day-to-day cares, we are not always consciously aware of it. Gazing up into the glorious spectacle of the night sky, we are confronted by the mystery of its origins and uncertain future. Astronomers gauge the universe to be a mind-boggling 13.8 billion years old. Similar estimations have dated our planet's origins to roughly 4.54 billion years ago. That original ball of cooling mass was undoubtedly lifeless. How astonishing that little more than half a billion years later living cells resembling bacteria appear in the fossil record. The evolution of those first cellular life forms is a second great mystery, wonderfully captured by my late friend Lynn Margulis and her son, Dorion Sagan, in their book, *Microcosmos*. It appears to have been a stage during which many of the metabolic pathways essential to cellular life were forged. But life cannot have begun with such complex organisms, with their cellular membranes and thousands of genes. It must surely have begun with much simpler entities that were closer in structure to viruses. The mystery then is how such prototypical entities might have evolved from inanimate chemicals at an astonishingly early stage of the evolving planet. If we wish to explore such a scenario we must look more closely at the basic nature and likely origins of viruses.

For all the harm they cause us, viruses are not evil. They do not think and they are incapable of feeling emotion. They are quintessentially amoral. But they are not free to do as they please. On the contrary, they are driven, and thus controlled, by definable and demonstrable evolutionary forces, which maximise their chances of survival, and through survival enable them to achieve maximum replicative success. The same evolutionary forces govern all of life on Earth. But in the case of viruses, these forces work much faster than they do on more complex cellular organisms. We have witnessed how viruses are symbiotic with every species of cellular life on Earth. That lightning-fast speed of evolution coupled with the fact that viruses replicate within the intimate landscape of the host cell, often the very genome, taking advantage of and potentially altering that genetic landscape, makes it inevitable that viruses will have influenced the evolution of cellular life from its very beginnings.

But what of the origins of viruses themselves?

Theories of the origins of viruses have varied over the span of little more than a century during which we have been aware of their existence. Even today, the true origin of viruses is unknown, leading to a variety of theories to explain their first beginnings. We can only extrapolate, from what we observe in their biological make-up, behaviour and properties today, how their forerunners might first have come into being. This has produced four basic theories for how viruses might have emerged.

The 'virus-first' theory proposes a primal origin in the prebiotic era of the Earth's evolution. The 'reduction' theory proposes a process of simplification and reduction from a previous unicellular stage of evolution. The 'escape' theory is really a development of the second, in proposing that viruses came about when a genomic fragment of a cellular life form, one that might have resembled the plasmid that is sometimes involved in genetic interchange between prokaryotic cells, escaped the control of its parental cells to become parasitically self-driven. The 'polyphyletic' theory takes into account the fact that viruses embody such an array of different genomic structures that they are most likely to have come into being through a variety of different origins. While acknowledging that all four theories have their pros and cons, and also while accepting that the most plausible explanation for the entire world of viruses is likely to have involved many different mechanisms over the vastness of evolutionary time, and thus the probability of polyphyletic origins for the wide range of variation we see in the diverse range of viruses today, I support a virus-first hypothesis for the origin of RNA viruses – and I propose that this was the primal origin of all viruses *per se* – in the stage of life known as the RNA World.

Much of the early theorising on the origin of viruses was influenced by the belief that viruses could not have evolved before cellular life existed, since viruses were defined as genetic parasites of cellular life. Even if we are to adopt a more comprehensive symbiotic interpretation of viruses, some evolutionary biologists will still insist that viruses could not have evolved before the evolution of cellular life forms, since, in their opinion, viruses can only exist where there are hosts for them to enter into symbiotic partnerships with. But I still take the view that we cannot make this assumption when it comes to RNA viruses. Viruses do not necessarily require cellular symbiotic partners. We have already witnessed examples of viruses partnering other viruses. I would suggest that there are other good reasons for believing that RNA viruses emerged from a prototypical RNA World.

There are quintessential differences between the chemical properties of the nucleotides DNA and RNA. Today we know that DNA is the molecule of heredity in all of cellular life, including humans. We have already come across a good reason for this: its chemical stability makes DNA the perfect storage vehicle for the genetic memory necessary for heredity to work. RNA is an entirely different 'kettle of fish'.

Darwin believed that natural selection must have operated from the very earliest stages of the evolutionary process. While the origin of cells, as the binding structures for cellular life, must have been a very important step in the evolution of cellular life, the origins of self-replicating nucleic acid chains as a source of both the genetic memory necessary for heredity and the coding for the proteins necessary for biological metabolism is an equally important step. Modern biochemical research confirms Darwin's belief that the established mechanisms of evolution can be extrapolated to the prebiotic stage, which is thought to have involved self-replicating polynucleotide chains. We also know that polynucleotide chains of DNA cannot self-replicate. DNA can only be copied with the help of protein-based enzymes known as DNA polymerases. RNA polynucleotide chains are capable of storing the genetic code, like DNA, meanwhile they are also capable of providing the catalytic, structural and regulatory roles necessary for self-replication. This has led leading chemists to propose that the prototypical stage of life is likely to have begun with RNA-based self-replicators, evolving within a prototypical world known as the RNA World.

If we now apply basic evolutionary theory to such self-replicating RNA chains, we find that mutations, arising through copying errors during replication, would have given rise to hereditable change in the sequences of daughter chains, just as it does in biotic evolution today. Alternatively, if two different pre-evolved chains coalesced to make a larger and more complex chain, this would result in a sudden increase in hereditary complexity as we see today with the fusion of genetic lineages in genetic symbioses. If we assume that Darwin was right, and natural selection operated at this primordial stage of evolution, the resultant mutants and holobionts would have competed with one another for survival in their primal world with the more successful self-replicators coming to dominate the local population. Scientific experiments have been set up to test these hypotheses and the outcomes have indeed turned out to be exactly as anticipated.

There are additional implications for the self-replicator stage of evolution that can be extrapolated from current theories of evolution, and which are specific to viral patterns of evolution.

When, in 1922, the German chemist and Nobel Laureate Manfred Eigen attempted to reproduce prebiotic evolution, he discovered that genetic self-replicators became parasitised by other self-replicators. This was the first confirmation that parasitic entities akin to viruses were not dependent on pre-existing cellular life forms. A generation or so later, the same phenomenon was observed by John von Neumann, who used computer modelling to create artificial life programs. His computer-based mathematical models were also parasitised by genetic self-replicators. This pattern of spontaneous parasitism has been further confirmed by experiments involving RNA-based viruses in cell cultures, and in more modern computer simulations. In all such scenarios parasitic elements have invaded, and then interacted with, the self-replicators. If further proof were necessary we have hard evidence from the modern world of virology, where key genes involved in viral replication, essential to both RNA- and DNA-based viruses, are not found in cellular life forms any more than the genes coding for the defining capsid proteins of DNA- and RNA-based viruses – the defining membranes for viruses that are the equivalent of the defining cell walls of cellular life.

Many authorities on the origins of life on Earth support the notion of a primal beginning in an RNA-based world. The virus-first model offers a logical basis for the origins of both RNA viruses and cellular life dating from that world. The subsequent evolutionary step from a genome based on RNA to one based on DNA would have required the replacement of a single nucleic acid in the chain, uracil, by thymine, which would have endowed a greater stability for the memory required for heredity across generations. That stability would, likely, have been selected for by natural selection. But in what primal landscape would such a primal interplay between self-replicators and natural selection be likely to have taken place?

In a letter to his mentor and friend, Joseph Hooker, dated 1871, Darwin wrote: *'But if (and oh what a big if) we could conceive in some warm little pond with all sorts of ammonia and phosphoric salts, light, heat, electricity etcetera present, that protein compound was chemically formed ready to undergo still more complex changes . . .'* It is a lovely image but I'm afraid that rather than Darwin's warm little pond, current opinion favours the broiling deep-sea hydrothermal vents as the likely source of life on Earth.

When biologists searched such seemingly hostile environments, at temperatures above a searing 80°C, they discovered a diversity of virus-like particles greatly exceeding the number found in aquatic systems at lower temperatures. These viruses appeared to thrive in such testing temperatures and violent surroundings. How remarkable it seems that such habitats might have provided prototypical

RNA self-replicators with the potential of furious ongoing evolutionary trial and error. Laboratory studies are now being conducted to examine the possibility of long-chain RNA evolution under similar environmental conditions to the hydrothermal vents. Such experiments suggest that naturally occurring mineral-rich surfaces, such as borates, apatite and calcite, may have helped to catalyse the formation of small organic compounds from inorganic compounds. These studies also confirm that RNA polynucleotides, the essential precursors to protobiotic RNA chains, can self-assemble in such intemperate conditions.

DNA, as we now know, is a highly stable molecule, which makes it ideal for the memory of genetic inheritance across generations. The sister molecule, RNA, as we also know, is much less stable. But in its essential instability, RNA may embody fast-moving evolutionary properties that make it likely, in the unstable environment of the hydrothermal vents, that RNA was the perfect molecule of heredity during the early stages of life's origins. Moreover, there is an additional observation that makes this even more likely to have been the molecule that kick-started the primal steps from chemical to biotic life.

The only organisms with RNA-coded genomes today are RNA-based viruses, suggesting that we might derive helpful insight into the purported RNA World from the study of RNA-based viruses. A key developmental step in the evolution from chemical self-replicators to life must have been the evolution of the concept of 'self'. A clue to such a primitive potential lies with this very capacity of RNA viruses, such as we saw with the example of HIV-1, to evolve as quasispecies. But what does this strange term really imply?

The term 'quasispecies' was introduced by the same pioneering German chemist, Manfred Eigen, when he was applying the

Darwinian concept of natural selection to the evolutionary behaviour of self-replicating polynucleotides. This concept proved useful to biologists who were studying the behaviour of RNA viruses in cultures and infected patients, where they witnessed how swarms of viruses, all closely related to one another through shared prior mutations, appeared to act as a single evolving entity in competition with other swarms and individual viruses for existence in a highly mutagenic environment. The evolution of quasispecies clusters appeared to confer a primal recognition of 'self' to the members of the swarms, giving them a measurable advantage for survival, even under conditions of extreme existential challenge. When quasispecies behaviour of RNA viruses was studied in various experimental situations, virologists noticed that even less-fit members of quasispecies would out-compete much fitter non-swarm rivals, confirming that natural selection was operating at the level of the swarm rather than the individual virus.

This pattern of RNA-mediated group identity was found to apply to many different experimental situations, whether involving tests on self-replicating polynucleotides or involving the actual behaviour of RNA viruses in the laboratory and inside infected patients. It supported the likelihood that RNA-virus-like entities played a key role as the primal recognition of 'self' in the origins of life during the presumptive RNA World stage. It also supported the theory of the origin of RNA-based viruses in the RNA World. The arrival of DNA as the molecule of heredity would have been a key stage in the origins of cellular organisms. The 'virus-first' perspective would have no difficulty in envisaging an origin and subsequent diversifying of DNA-based viruses from RNA-based precursors, involving multiple genetic symbiotic exchanges between viruses and hosts. The ease with which both RNA- and DNA-based viruses establish a great variety of genetic symbioses with all

DNA-based cellular organisms today supports the likelihood of such ongoing genetic interactions. That continuing interaction of both RNA- and DNA-based viruses with all evolving cellular life forms would readily explain the evolution of biodiversity and the complex interactive ecologies we witness today.

23

The Fourth Domain?

For much of the twentieth century, biologists were agreed that life was defined as five kingdoms, which included the animals, the plants, the fungi, the single-celled nucleated protists and the bacteria. This definition assumed that life was exclusively cellular. Moreover, the differences between these five cellular kingdoms required little more than the aid of the ordinary lab microscope. There was, of course, a major distinction between the first four kingdoms and the bacteria: the animals, plants, fungi and protists that comprised the eukaryotes had cells in which the genomes were walled off within a central compartment, called the nucleus, meanwhile the bacteria, or prokaryotes, comprised single living cells containing a ring genome without any separately compart-mentalised nucleus. This system of classification remained the mainstay of biological classification for almost a century. Then, out of the blue in 1977, an American microbiologist, Carl Richard Woese, working with fellow microbiologist Ralph S. Wolfe in the Microbiology faculty at the University of Illinois at Urbana-Champaign, contradicted this classification in an iconoclastic paper published in the *Proceedings of the National Academy of Sciences of the United States of America*. In this, and subsequent, papers Woese began to dismantle the Five Kingdom classification and replace it with a radical alternative.

To begin with they proposed that the prokaryotes could not be designated a single kingdom, but needed to be separated into two basically different biological 'domains'. The two domains comprised the 'Eubacteria' – the familiar bacteria, such as caused tuberculosis, or the *E. coli* that resided in the human colon – and what Woese initially termed the 'Archaebacteria'. Even as his proposal was still proving highly controversial among evolutionary biologists, Woese abandoned his term Archaebacteria and replaced it with the simpler term, 'Archaea', which comes from the Greek for 'ancient things'. In his concept and definition of Archaea, Woese believed that he had discovered not just a new division of bacteria but a completely new domain of life.

Archaea, Woese now attested, should be seen as the earliest form of cellular life on Earth, inhabiting primitive anaerobic ecologies – this explained why their internal chemistry made use of such primal chemicals as methane and hydrogen sulphide. He further proposed that the long-established 'tree of life' needed to be torn from its roots and reclassified into three distinct domains: the Archaea, the Eubacteria and the Eukaryotes – this latter group embracing all varieties of cellular life other than Archaea and Eubacteria, thus including the animals, plants, fungi and the single-celled nucleated protists such as the amoeba.

What had driven Woese to such iconoclastic conclusions?

To understand his reasoning, we need to grasp that Woese was looking not at the diversity of life as we see it today, but life back at the time of its very origins – the stage of single-celled ancestors some billions of years before the present. Life at this stage would have been exclusively microbial and thus unlikely to leave much in the way of clues in the fossil record. This obliged him to discover new ways of interpreting this primal stage of the evolution of life. In 1997 Woese explained his basic line of thinking: 'We had a real evolutionary understanding of the plants and animals, but that left out the whole world of the bacteria. So I thought that's what I would do first: bring in the prokaryotes.'

With no clues to be found in the fossil record, Woese focused instead on the genetic and biochemical record that was there to be examined in the most basic chemistry within present-day living cells. He also believed that life must have begun with RNA moieties. In particular he focused on the RNA-based molecules within the cytoplasmic structures called the ribosomes: these are the protein-manufacturing factories of every living cell. Reasoning that protein-manufacturing processes must be of very ancient origins, he trusted these to provide him with the perfect tool for exploring patterns of evolutionary descent over the vastness of time, extending to billions of years.

When Woese compared ribosomal RNAs between different groups of what were then regarded as bacteria, he made a major discovery. The ribosomal RNAs of bacteria were not always the same. There was a group of bacteria, known as methanogens, that, when viewed down the microscope, looked exactly like the other bacteria – but their ribosomal RNA sequences were very different. The ability to metabolise methane suggested a very primitive origin. These methanogens were the bacteria that he initially renamed Archaebacteria. But as he continued to study them it dawned on him that they were just too different biochemically to be closely related in their evolutionary origins to the majority of familiar bacteria. He concluded that these Archaebacteria must have come from a different evolutionary lineage from that of the bacteria, a difference that must date back to the very beginnings of cellular existence. Since they also had features that suggested more ancient origins than the more familiar bacteria, which he now called the Eubacteria, he renamed the Archaebacteria as simply 'Archaea'.

It was inevitable that traditional biologists would be sceptical of Woese's shake-up of the fundamental roots of the phylogenetic tree. In Woese's classification, Eubacteria and Archaea were more different from one another than, say, an amoeba and an oak tree. Woese's ideas were subjected to brutal criticism from many quarters, including some of the most distinguished evolutionary biologists in the world.

Acceptance of his reclassification was not helped by the fact that Woese was a shy and retiring individual, with an aversion to attending scientific meetings. But he refused to be cowed and continued to explore the ramifications of his discovery, never doubting its veracity and implications. Meanwhile, as the range of subtle, but vital, differences between Archaea, Eubacteria and Eukaryotes became more widely appreciated by an increasingly genetically-focused generation of biologists, the logic of 'the Woesian revolution' began to win through. Today most evolutionary biologists no longer classify life into the former five kingdoms but into Woese's three domains. Within this classification, the domain of the Eukaryotes, or 'true nucleated life forms', now includes all of the first four kingdoms of the former classification: animals, plants, fungi and single-celled nucleated protists. Meanwhile, the biological world agrees that the Eubacteria and the Archaea are separate domains.

Broadly speaking, Eubacteria have more complex genomes, and internal chemistry, than Archaea. They are also much more widely distributed so that, when biologists loosely talk about bacteria, they are usually referring to Eubacteria. Archaea differ from Eubacteria in the chemistry of their cell walls, and also in the chemical structures of several key enzymes involved in DNA replication and its transcription into messenger RNA and subsequent translation into proteins. All of this confirms Woese's way of thinking, in that the Archaea really have key features that suggest they are descended from the earliest cellular organisms, evolving at a time when the Earth was less hospitable than it is today. Eubacteria, on the other hand, can live in both an oxygenated and anaerobic environment. They are found in most of the common aqueous and terrestrial ecologies: meanwhile Archaea inhabit ecologies that tend to be relatively harsh and devoid of oxygen.

Viruses, being non-cellular, did not figure in Woese's domains, any more than they figured in the former divisions of the five-kingdom classification. But, as we have seen in the previous chapters, viruses have played an enormously important role in the origins of life and its subsequent diversification. How then should we interpret viruses and their role in the evolution of life? Any such interpretation will inevitably be speculative, but perhaps our first step should be to look closely at viruses, paying particular attention to their genetic and biochemical make-up, and their familiar patterns of behaviour, without any preconceived notions.

Let us start by asking the question that recently provoked controversy in microbiological journals: are viruses a fourth domain of cellular life?

In my opinion the answer is an unequivocal 'No!' Viruses are not cellular. They do not have cellular membranes and they do not contain the typical genetic and biochemical features of cellular organisms – for example, they lack the very ribosomes that Woese focuses on in his basic approach to cellular evolution.

Let us then pose a new question: what structures do viruses have that cellular organisms do not have?

The answer is that while viruses do not have a cell membrane, they do possess an alternative membrane only found in viruses. We have encountered it again and again in previous chapters of this book. It is, of course, the capsid. So now we have pinned down two characteristics that are quintessential to viruses and to none of the domains of cellular life: they are non-cellular and their genome codes for the quintessential viral structures that enclose every viral genome: it is called a 'capsid'.

Let us move on to consider genomes. There is no structure more important to the evolution of life than the genome that both stores and codes for its physical make-up and inheritance. We have seen how viruses also possess genomes, albeit these are usually much smaller and more compact than the genomes of cellular organisms. So the possession of a genome is an important part of the argument for viruses being considered as life forms. Moreover, there is a vital clue here that also sets viruses apart from all three cellular domains. This is the fact that some viruses – a minority of those that exist today, but a considerable minority nevertheless – have genomes based not on the DNA that codes for all cellular life forms, but based on RNA. If life truly began with the key stage of an RNA-based world, there is a potential link here to the evolutionary origins of RNA-based viruses. With this comes an extraordinary potential of acquiring a primal establishment of the concept of 'self', which is an integral requirement for one of the key steps in the evolution from self-replicating polynucleotides to the primal origins of life.

Let us pose another question: is there evidence for the existence of viruses during the most ancient phase of the evolution of cellular life?

In 1974 a virus was found to 'infect' an archaean. This proved to be a 'halovirus' – a virus that could survive extremes of salinity. Since its discovery preceded Woese's reclassification, it was initially classed as a bacteriophage of what was then assumed to be a bacterial host, mistakenly named *Halobacterium salinarum*. Today the halovirus would be reclassed as an archaeal virus. This was followed by a series of discoveries in relation to archaeal viruses. One such discovery was a novel mechanism of release of the daughter viruses, through pyramidal structures, sometimes constructed with icosahedral symmetry, on the cell membrane of the archaeal host. No such mechanism has ever been seen in the viruses in any other domain. The first virus isolated from a methanogen was reported in 1986. These somewhat strange viruses now followed a pattern, which included difficulties in isolation, and thus study, of the viruses, because of difficulties in culturing their host Archaea. The first thermophilic viruses were isolated from sulphur-dependent hosts in the early 1980s. Since then more than 117 archaeal viruses have been identified from a diverse range of archaeal hosts and from a wide range of environments that include both extreme and non-extreme environments. Experts in the field think that we have just begun to discover what will ultimately prove to be a major diversity of archaeal viruses in the biosphere. The same experts stress that although the Archaea remain the most enigmatic of life's three domains, the diversity of their virion shapes and the archaeal virus genome variations is already proving remarkable. This became evident when some 29 archaeal viruses were newly reclassified, when they were found to represent some 15 different viral families. All 6,000 known eubacterial viruses are classed as belonging to just 10 families. This suggests that the viruses of Archaea are more ancient than the viruses of the other two domains and are likely to show much wider genetic variation.

The archaeal viruses are indeed proving to be stranger than any other virus group studied to date, some being shaped like bottles, or spindles, with short and long tail-like appendages, others being droplet-shaped with beard-like fibres, or shaped like coils or spheres, some being capable of shooting out new tails after they have exited their hosts. The complexity of what these viruses are doing, in symbiotic interplay with their archaeal hosts, is illustrated by a detailed study over many years involving a single acidic hot spring within Yellowstone National Park.

As might be expected, the microbial community in this harsh and primitive environment was relatively modest, with 97 per cent of the cells classed as Archaea, 3 per cent Eubacteria and no Eukarya. This microbial community appeared to be relatively stable over several years. The viral component was clearly dominated by archaeal viruses. Interestingly, the majority of these archaeal viruses were RNA-based. But these viral RNA genomes lacked a key enzyme involved in the replication mechanisms of the RNA-based viruses of eukaryotic or eubacterial hosts, once again suggesting that these might be the most ancient of all known viruses. The microbiologists, studying what may well be a microcosm of the earliest stage of life on Earth, concluded that what they discovered clearly emphasised 'the central role that viruses play in causing disease, controlling microbial community composition and structure, and driving evolution'.

The biological world is changing its former perspective of viruses. At the very base of the biological food chain in important ecologies, viruses are 'managing' the microbial communities, and through a series of aggressive symbiotic interactions, they are stabilising the ecologies. These findings of the importance of the viral contribution to the deep levels of ecological balance, coupled with the numerous examples of viruses interacting at aggressive symbiotic level with all three cellular domains of life, including the mammalian branch of the eukaryotes, point to a vital contribution of viruses to biodiversity.

This brings back into focus the controversial role of viruses in relation to the evolutionary tree of life.

We humans are apt to regard the Earth as 'our world', assuming that we have hegemony over it. But the truth is that we are hardly vital to biodiversity. Our burgeoning population, with its increasing intrusion into wilderness areas, destruction of rain forests and over-fishing of the oceans, is putting a strain on the natural balances of a number of major ecologies, meanwhile contributing to the extinction of many other forms of life. We inhabit a planet only recently recovered from the ozone crisis, meanwhile we face possible climate change and the plastic pollution of the oceans. It is salutary to reflect on the fact that we inhabit a virosphere that teems with unknown viruses. Putting aside the threat of emerging diseases from viruses, those same, amoral entities have played a key role in the origins of life on Earth and in its evolution to biodiversity as we know it. If we are inclined to doubt the creative role of viruses to biodiversity, we might ask ourselves a single salient question: what would happen if viruses were to disappear from the face of the Earth? We only need to consider those gargantuan cycles of nutrients in the oceans, and perhaps in major land ecologies, to hazard a guess at the answer.

Those who refute the notion of viruses as living organisms point to the fact that they are not capable of replication by themselves, and so dismiss them as living organisms. But this is to misinterpret the essential nature of symbiogenesis. Every virus has evolved as a symbiotic partner with its cellular host: it is in the very nature of this existential interaction that viruses depend on the host for replication. In return, and through sharing the hosts' life cycles, viruses have massively contributed to the hosts' evolution. The key observation, that all viruses are obligate symbionts, is the final ingredient of my definition of viruses as living entities. This, if I might remind readers, is why I proposed a new definition of viruses, taking all that we have learned into consideration:

Viruses are non-cellular, capsid-encoding, obligate symbionts.

After much consideration, I suggest that RNA viruses originated as symbionts of the evolving RNA-based self-replicators in the RNA World. Then, as life subsequently evolved to the three cellular domains, viruses continued to evolve and diversify, in symbiotic partnership with the cellular domains, playing a highly interactive and creative role in the origins and diversification of the cellular tree of life. That role continues on a global scale even today. This makes it somewhat specious to separate viruses, in the minds of biologists, from the cellular domains with which, in the words of Durzyńska and Goździcka-Józefiak, 'viruses have been entwined since the dawn of evolution'.

In practical terms the discipline of virology has developed, much as viruses have themselves evolved, as a separate yet intertwined field of study, with the broader field of biology. This is increasingly outdated as evolutionary biologists and ecologists are coming to understand the vast weave of interconnections between viruses and the cellular domains. I suggest that there is now an over-whelming argument for viruses to be assigned their own biological domain, whether this is termed the 'Fourth Domain of life', or perhaps simply, 'The Viral Domain'.

Bibliography and References

A more scientific reader may find this guide to further reading useful. I have listed a few relevant books at the beginning. A small number of references may recur in different chapters because they cover more than a single theme. Readers may also obtain useful references at my website www.fprbooks.com.

Books

Collier L. and Oxford J., *Human Virology*. Oxford University Press, 1993.

Field B.N. and Knipe D.M., *Field's Virology*. Raven Press, New York, 1990.

Margulis L. and Sagan D., *Microcosmos: Four Billion Years of Microbial Evolution*. University of California Press, Berkeley, Los Angeles, London, paperback, 1997.

McNeill W.H., *Plagues and Peoples*. Basil Blackwell, Oxford, 1977.

Nibali L. and Henderson B., eds, *The Human Microbiota and Chronic Disease*. Wiley Blackwell, Hoboken New Jersey, 2016.

Ryan F., *Virus X*. Little Brown and Company, Boston, New York, Toronto and London, 1997.

Ryan F., *Virolution*. HarperCollins Publishers Ltd, London, 2009.

Ryan F., *The Mysterious World of the Human Genome*. HarperCollins Publishers Ltd, London, 2015.

Summers W.C., *Félix d'Herelle and the Origins of Molecular Biology*. Yale University Press, 1999.

Villarreal L.P., *Viruses and the Evolution of Life*. ASM Press, Washington D.C., 2005.

Epigraph

Anthony Hopkins Interview in *The Sunday Times* Colour Supplement, 12 April 1992.

Introduction

The book on the human genome: see Ryan F., *The Mysterious World of the Human Genome*, in Books.

Chapter 2

For more details of the human microbial flora: see Nibali and Henderson in books above.

Chapter 3

Hankin E.H., L'action bactéricide des eaux de la Jumna et du Gange sur le vibrion du choléra. *Annales de l'Instituté Pasteur*, 1896; **10**: 511-523.

Twort F.W., An investigation on the Nature of Ultra-Microscopic Viruses. *The Lancet*, 1915; **186**: 4814.

D'Hérelle Félix, Sur un microbe invisible antagoniste des bacilles dysentériques. *Comptes Rendus de l'Adadémie des Sciences de Paris*, 1917; **165**: 373-375.

D'Herelle's references to bacteriophages as symbionts, comparing them to the mycorrhize of orchids: D'Herelle F., *The Bacteriophage and Its Behaviour*. Ballière, Tindall and Cox, London, 1926. Chapter V: p.211. (NB on p.343 d'Herelle defends the bacteriophage as living. See also pp.326 and 354.)

Chapter 4

WHO figures and advice on measles: www.who.int/news-room/fact-sheets/detail/measles.

'Measles rise worldwide from 2017 to 2018'. *New Scientist*, 24 February 2018, pp.4–5.

'Measles is back with a vengeance – is the anti-vaccination movement to blame?' Chloe Lambert, *Daily Telegraph*, 7 May 2018.

For GPs put on alert over surge in measles: Chris Smyth, *The Times*, 3 July 2018.

Rubella and links to teratogenicity: Lee J-Y, and Bowden D.S., Rubella Virus Replication and Links to Teratogenicity. *Clin. Microbiol. Rev.*, 2000; **13**(4): 571-587.

Chapter 5

How noroviruses cause disease: Karst S.M., Pathogenesis of Noroviruses, Emerging RNA Viruses. *Viruses*, 2010; **2**: 748-781. See also: Karst S.M. and Wobus C.R. A Working Model of How Noroviruses Infect the Intestine. *PLOS Pathogens*, February 27, 2015 | doi:10.1371/journal.ppat.1004628.

Chapter 6

For more details of Franklin D. Roosevelt: see FDR Presidential Library & Museum online.

Chapter 7

The role of smallpox in the European conquest of the Americas: See McNeill W.H. in recommended books.

Smallpox virus's inhibition of interferon: Del Mar M. and de Marco F., The highly virulent variola and monkeypox viruses express secreted inhibitors of type I interferon. *FASEB J.*, 2010; **24**(5): 1479-1488.

Chapter 8

For a more detailed contemporary description of the *Sin Nombre* hantavirus outbreak, see *Virus X* in books.

Chapter 9

Furman D., Jolic V., Sharma S., et al., Cytomegalovirus infection enhances the immune response to influenza. *Sci. Translational Med.*, 2015; 7(281): doi 1-.1126/scitranslmed.aaa.2293.

Reese T.A., Co-infections: Another Variable in the Herpesvirus Latency-Reactivation Dynamic. *J. Virol.*, 2016; doi 10.1128/JVI.01865-15.

Cytomegalovirus frequency in US populations: Staras S.A., Dollard S.C., Radford K.W., et al., Seroprevalence of cytomegalovirus infection in the United States, 1988-1994. *Clin. Infect. Dis.*, 2006; 43(9): 1143-1151.

Butkitt's paper on lymphoma in African children: Burkitt D., A sarcoma involving the jaws in African children. *Br. J. Surg.*, 1958; 46: 218.

Chapter 10

Deaths from influenza during World War I: Wever P.C. and van Bergen L., Death from 1918 pandemic influenza during the First World War: a perspective from personal and anecdotal evidence. *Influenza and Other Respiratory Viruses*, 2014; 8(5): 538-546. doi:10.1111/irv.12267.

Information on SARS: Smith R.D., Responding to global infectious disease outbreaks. Lessons from SARS on the role of risk perception, communication and management. *Social Science and Medicine*, 2006; 63(12): 3113-3123.

Bird Flu 2017 in China: MacKenzie D., Lethal flu two genes away. *New Scientist*, 24 June 2017: 22-23.

Chapter 11

The rabies case report: McDermid R.C., Lee B., et al., Human rabies encephalitis following bat exposure: failure of therapeutic coma. *C.M.J.*, 2008; **178**(5): 557-561.

The lesson of the myxomatosis virus and the Australian rabbit: Kerr P.J., Liu J., Cattadori I., et al., Myxoma Virus and the Leporipoxviruses: An Evolutionary Paradigm. *Viruses*, 2015; 7: 1020-1061. doi:10.3390/v7031020.

Chapter 12

For a detailed contemporary description of the initial Ebola outbreak: see *Virus X*.

For the story of the 2014 West African outbreak, and most particularly, the neurological complications: Billioux B.J., Smith B. and Nath A., Neurological Complications of Ebola Virus Infection. *Neurotherapeutics*, 2016; **13**: 461-470.

Bats as source of viruses: Olival K.J. and Hayman D.T.S., Filoviruses in Bats: Current Knowledge and Future Directions. *Viruses*, 2014; **6**: 1759-1788.

Bats as source of other viruses: Marsh G.A., de Jong C., Barr J.A., et al., Cedar Virus: A Novel Henipavirus Isolated from Australian Bats. *PLOS Pathogens*, 2012; **8**(8): e1002836. See also: Olival K.J., Hosseini P.R., Zambrana-Torrelio C., et al., Host and viral traits predict zoonotic spillover from mammals. *Nature*, 2017: **546**: 646-650.

Chapter 13

Zika and brain complications: Da Silva I.R., Frontera J.A. and
Bispo de Filippis A.M., Neurologic Complications Associated
with the Zika Virus in Brazilian Adults. *JAMA Neurol*, 2017;
doi:10.1001/namaneurol.2017.1703.

The use of *Wolbachia* in mosquito control. *Daily Telegraph*, UK,
2016/10/26/infected mosquitoes-to-be-released-in-Brazil-and-
Columbia . . .

Chapter 14

For details of the history of hepatitis: Trepo C., A brief history
of hepatitis milestones. *Liver International*, 2014. doi.10.1111/
liv.12409.

The WHO statistics for hepatitis B: www.who.int/news-room/fact-
sheets/detail/hepatitis-b.

Information on rising incidence of hepatitis E in UK: see UK.
gov website.

Chapter 15

The Cromwell quote: Burns D.A., 'Warts and all' – the history
and folklore of warts: a review. *J Roy Soc Med*, 1992; **85**:
37-40.

The zur Hausen note: Zur Hausen H., Condylomata Acuminata
and Human Genital Cancer. *Cancer Research*, 1976; **36**: 794.

The Who Bulletin on the incidence of cervical cancer and HPV
vaccines: Cutts F.T., Franceschi S., Goldie S., et al., Human
papillomavirus and HPV vaccines: a review. 2007. www.who.
int/vaccines-documents/DocsPDF07/866.pdf.

HPV vaccination in England: Johnson H.C., Lafferty E.I., Eggo R.M., et al., Effect of HPV vaccination and cervical cancer screening in England by ethnicity: a modelling study. *The Lancet*, 2018; 3:e44-51. http://dx.doi.org/10.1016/s2468.2667(17)30238-4.

HPV vaccination in Scotland: Narwan, G. Vaccine drive cuts cancer virus by 90 per cent. *The Times*, 6 April 2017.

HPV vaccine in Ireland: Coyne, Ellen, Senior Ireland Reporter, *The Times*, August 2 2017.

HPV in the United States: The Henry Kaiser Family Foundation, October 2017 Factsheet. HPV Vaccine: Access and Use in the US.

Chapter 16

Mimivirus genome: Raoult D., Audic S., Robert C., et al., The 1.2-megabase genome sequence of Mimivirus. *Science*, 2004; **306**: 1344-1350.

Breaking epistemological barriers: Claverie J-M and Abergel C., Giant viruses: the difficult breaking of multiple epistemological barriers. *Studies in History and Philosophy of Biological and Biomedical Sciences*, 2016; **59**: 89-99.

Klosneuvirus: Schulz F., Yutin N., Ivanova N.N., et al., Giant viruses with an expanded complement of translation system components. *Science*, 2017; **356**: 82-85.

Sequences similar to Mimivirus in the Sargasso Sea: Ghedin E. and Claverie J-M, Mimivirus Relatives in the Sargasso Sea. *Virol. J.*, **2**: 62. doi:10.1186/1743-422X-262.

Curtis Suttle quote: Science Daily, 2011. World's Largest, Most Complex Marine Virus Is Major Player in Ocean Ecosystems. www.sciencedaily.com/releases/2010/10/101025152251.htm.

Conflicts involving viral giants: Forterre P., Giant Viruses:

Conflicts in Revisiting the Virus Concept. *Intervirology*, 2010; **53**: 362-378.

Giant viruses in Antarctica: Kerepesi C. and Grolmusz V., The 'Giant Virus Finder' discovers an abundance of giant viruses in the Antarctic dry valleys. *Arch Virol.*, 2017; **162**: 1671-1676.

Origins of giant viruses from smaller viral predecessors: Yutin N., Wolf Y.I. and Koonin E.V., Origin of giant viruses from smaller DNA viruses not from a fourth domain of cellular life. *Virology*, 2014; **466-467**: 38-52.

Definition of viruses as capsid-encoding organisms: Forterre P. and Prangishvili D., The great billion-year war between ribosome- and capsid-encoding organisms (cells and viruses) as the major source of evolutionary novelties. *Ann. N.Y. Acad. Sci.*, 2009; **1178**: 65-77.

The four mechanisms of the acronym MESH: Ryan F.P., Genomic creativity and natural selection: a modern synthesis. *Biological Journal of the Linnean Society*, 2006; **88**: 655–672.

Chapter 17

Excluding viruses from the tree of life: Moreira D. and López-Garcia P., Ten reasons to exclude viruses from the tree of life. *Nature Reviews|Microbiology*, 2009; 7: 305-311.

Key genes for proteins involved in viral replication only found in viruses: Koonin E.V., Senkevich T.G. and Dolja V.V., 2006. The ancient Virus World and the evolution of cells. *Biology Direct*. doi:10.1186/1745-6150-1-29.

Genes encoding the major capsid proteins on found in viral genomes: Prangishvili D. and Garrett R.A., 2004. Exceptionally diverse morphotypes and genomes of crenarcheal hyperthermophilic viruses. *Biochem. Soc. Trans.* **32**(2): 204-208. See also Koonin, Senkevich and Dolja 2006.

Retroviruses and bacteriophages not originating from cellular offshoots: Villarreal L.P., 2007. Virus–host symbiosis mediated by persistence. *Symbiosis*. **44**: 1-9. See also Hambly E. and Suttle C.A., 2005. The virosphere, diversity, and genetic exchange within phage communities. *Curr Opinion Microbiol*. **8**: 444-450.

Viruses and all three cellular domains intertwined in their evolution: Durzyńska J. and Goździcka-Józefiak A. Viruses and cells intertwined since the dawn of evolution. *Virol. J.*, 2015; **12**: 169. doi: 10.1186/s12985-015-0400-7.

Chapter 18

All polydnaviruses from a single source: Provost B., Varricchio P. and Arana E., et al., Bracoviruses contain a large multigene family coding for protein tyrosine phosphatases. *J. Virol.*, 2004; **130**: 90-103.

Single unique origin to the wasp–virus symbiosis: Whitfield J.B., Estimating the age of the polydnavirus/braconid wasp symbiosis. *Proc. Natl. Acad. Sci. USA*, 2002; **99**(11): 7508-7513. See also Belle E., Beckage N.E., Rousselet J., et al., Visualization of polydnavirus sequences in a parasitoid wasp chromosome. *J. Virol.*, 2002; **76**: 5793-5796.

Chapter 19

The Suttle quote: Suttle C.A., Viruses in the sea. *Nature*, 2005; **437**: 356-361.

Some other papers related to the oceanic virosphere: Danovaro R., Dell'Anno A., Corinaldesi C., et al., Major viral impact on the functioning of the benthic deep-sea ecosystems. *Nature*, 2008; **454**: 1084-1087. Mulkidjanian A.Y., Koonin E.V., Makarova K.S., et al., The cyanobacterial genome core and the origin of photosynthesxis. *P.N.A.S.*, 2006; **103**(35): 13126-13131. Lindell D., Sullivan M.B., Johnson Z.I., et al., Transfer of photosynthesis genes to and from Prochlorococcus viruses. *P.N.A.S.*, 2004; **101**(30): 11013-11018.

Phage viruses 'a major component' of the oceanic environment: Krupovic M., Prangishvili D., Hendrix R.W. and Bamford D.H., Genomics of Bacterial and Archaeal Viruses: Dynamics within the Prokaryotic Virosphere. *Microbiol. and Mol. Biol. Rev.*, 2011; **75**(4): 610-635.

The great virus comeback: Forterre P., *The Great Virus Comeback* (translated from the French). *Biol. Aujourdhui*, 2013; **207**(3): 153-168.

Viruses more numerous than every other organism put together, including the bacteria, by an order of ten to a hundred-fold: Koonin E.V. and Dolja V.V., A virocentric perspective on the evolution of life. *Curr. Opin. Virol.*, 2013; **3**(5): 546-557.

Global genetic diversity of viruses: Angly F.E., Felts B., Breitbart M., et al., The Marine Viromes of Four Oceanic Regions. *PLOS Biology*, 2006; **4**(11): 2121-2131.

Viruses as drivers of global geochemical cycles: See Suttle 2005 above; also Rosario K. and Breitbart M. Exploring the viral world through metagenomics. *Curr. Opin. Virol.*, 2011; **1**(1): 289-297.

Chapter 20

Marilyn Roossinck's review paper: Roossinck M.J., Symbiosis versus competition in plant virus evolution. *Nature Rev. Microbiol.*, 2005; **3**: 917-924.

The paper on the three-way virus-infecting-fungus-infecting-plant: Márquez L.M., Redman R.S., Rodriguez R.J. and Roossinck MJ., A Virus in a Fungus in a Plant: Three-Way Symbiosis Required for Thermal Tolerance. *Science*, 2007; **315**: 513-515.

Four different fungus-protecting viruses: Xu P., Chen F., Mannas J.P., et al., Virus infection improves drought tolerance. *New Phytologist*, 2008; doi: 10.1111/j.1469-8137.2008.02627.x.

The prokaryotic virosphere: Krupovic M., Prangishvili D., Hendrix R.W. and Bamford D.H., Genomics of Bacterial and Archaeal Viruses: Dynamics within the Prokaryotic Virosphere. *Microbiol. and Mol. Biol. Rev.*, 2011; **75**(4): 610-635.

Viruses in six different ecologies of Delaware soils: Williamson K.E., Radosevich M. and Wommack K.E., Abundance and Diversity of Viruses in Six Delaware Soils. *Appl. Environ. Microbiol.*, 2005; **71**(6): 3119-31125.

Viruses in the cryptic Antarctic soils: Williamson K.E., Radosevich M., Smith D.W. and Wommack K.E., Incidence of lysogeny within temperate and extreme soil environments. *Environ. Microbiol.*, 2007; **9**: 2563-2574.

Viruses in coastal environments: Srinivasiah S., Bhavsar J., Thapar K., et al., Phages across the biosphere: contrasts of viruses in soil and aquatic environments. *Res Microbiol.*,2008; **159**: 349-357.

Williamson K.E, Fuhrmann J.J., Wommack K.E. and Radosevich M., Viruses in Soil Ecosystems: An Unknown Quantity

Within an Unexplored Territory. *Ann. Rev. Virol.*, 2017; **4**: 201-219.

Need for plant metagenomic studies: Roossinck M.J., Martin D.P. and Roumagnac P., Plant Virus Metagenomics: Advances in Virus Discovery. *Phytopath. Rev.*, 2015; **105**: 716-727.

Soil studies extended to Kogelberg Reserve in the Cape of South Africa: Segobola J., Adriaenssens E., Tsekoa T., et al., Exploring Viral Diversity in a Unique South African Soil Habitat. *Sci. Reports*, 2018; doi:10.1038/s41598-017-18461-0.

Soil studies further extended to Peru, California desert, Kansas prairie and paddyfields of Japan and Korea: Rosario K. and Breitbart M., Exporing the viral world through metagenomics. *Curr. Opin. Virol.*, 2011; **1**: 289-297.

Abundance of viruses in hydrothermal vents: Prangishvili D. and Garrett R.A., Exceptionally diverse morphotypes and genomes of crenarchaeal hyperthermophilic viruses. *Biochem. Soc. Trans.*, 2004; **32**(2): 204-208.

Implications of human virome for transplantation: Tan S.K., Relman D.A. and Pinsky B.A., The Human Virome: Implications for Clinical Practice in Transplantation Medicine. *J. Clin. Microbiol.*, 2017; **55**(10): 2884-2893.

Virosphere of the human gut: Aggarwala V., Liang G. and Bushman D., Viral communities of the human gut: metagenomic analysis of composition and dynamics. *Mobile DNA*, 2017; **8**:12. doi 10.1186/s13100-017-0095-y.

De la Cruz Peña M.J., Martinez-Hernandez F., Garcia-Heredia I., et al., Deciphering the Human Virome with Single-Virus Genomics and Metagenomics. *Viruses*, 2018, *10*, 113; doi.10.3390/v10030113.

Unknown virus in majority of metagenomic studies of gut virome: Dutilh B.E., Cassman N., McNair K., et al., A highly abundant bacteriophage discovered in the unknown sequences of human faecal metagenomes. *Nat. Comms.*, 2014|5:4498| doi:10.1038/ncomms5498|www.nature.com/naturecommunicationsarticles.

Chapter 21

The detailed story of the discovery of the HIV-1 virus is told in Chapter 13 of *Virus X*.

Endogenous retroviruses in amphibians and fish, shark and frog: Aiewsakun P and Katzourakis A., Marine origin of retroviruses in the early Palaeozoic Era. *Nature Comms.*, 2017. doi: 10.1038/ncomms13954.

Retroviruses in the photosynthetic sea slug, *Elysia chorotica*: Pierce S.K., Mahadevan P., Massey S.E., et al., A Preliminary Molecular and Phylogenetic Analysis of the Genome of a Novel Endogenous Retrovirus in the Sea Slug *Elysia chlorotica. Biol. Bull.*, 2016; **231**: 236-44.

The role of HERVs in embryological development, immunology and cellular physiology: See Villarreal 2005 in books; see also Ryan F.P., Viral symbiosis and the holobiontic nature of the human genome. *APMIS* 2016; **124**: 11-19.

The discovery of syncytin-1: Mi S., Lee X. and Li X., et al., Syncytin is a captive retroviral envelope protein involved in human placental morphogenesis. *Nature*, 2000; **403**: 785-789; Mallet F., Bouton O., Prudhomme S., et al., The endogenous retroviral locus ERVWE1 is a bona fide gene involved in hominoid placental physiology. *Proc. Natl Acad. Sci. USA*, 2004; **101**: 1731-1736.

The discovery of syncytin-2: Blaise S., de Parseval N., Bénit L., et al., 2003. Genomewide screening for fusogenic human endogenous retrovirus envelopes identifies syncytin 2, a gene conserved on primate evolution. *Proc. Natl Acad. Sci. USA,* 2003; **100**: 13013-13018.

Twelve viral loci involved in human reproduction: Villarreal L.P. and Ryan F., Viruses in host evolution: general principles and future extrapolations. *Curr. Topics in Virol.,* 2011; **9**: 79-90.

The role of syncytins and other endogenous retroviral genes in human placental abnormalities: Bolze P.A., Mommert M. and Mallet F., Contribution of Syncytins and Other Endogenous Retroviral Envelopes to Human Placental Pathologies. *Progress in Mol Biol and Transl Sci.,* 2018. In press.

The contribution of viruses in the autoimmune disorders and cancer: Ryan F.P., An alternative approach to medical genetics based on modern evolutionary biology. Part 3: HERVs in disease. *J. Royal Soc. Med.,* 2009; **102**: 415-424; Ryan F.P., An alternative approach to medical genetics based on modern evolutionary biology. Part 4: HERVs in cancer. *J. Royal Soc. Med.,* 2009; **102**: 474-480.

The syncytins in the many different mammalian families: Cornelis G., Heidmann O., Bernard-Stoecklin S., et al., Ancestral capture of syncytin-*Car1,* a fusogenic endogenous retroviral envelope gene involved in placentation and conserved in Carnivora. *Proc. Natl. Acad. Sci. USA,* 201; **109**(7): www.pnas.org/cgi/doi/10.1073/pnas.1115346109; Cornelis G., Heidmann O., Degrelle S.A., et al., Captured retroviral envelope syncytin gene associated with the unique placental structure of higher ruminants. *Proc. Natl, Acad. Sci. USA,* 2013. www.pnas.org/cgi/doi/10.1073/pnas.1215787110; Cornelis G., Vernochet C., Malicorne S., et al., Retroviral envelope syncytin capture in an ancestrally diverged mammalian clade for placentation in the primitive Afrotherian tenrecs. *Proc. Natl Acad. Sci. USA,* 2014; www.pnas.org/cgi/doi/10.1073/pnas.1412268111.

Retroviruses in the origins of the placental mammals: Cornelis
G., Vernochet C., Carradec Q., et al., Retroviral envelope gene
captures and syncytin exaptation for placentation in marsupials.
Proc. Natl. Acad. Sci. USA, 2015; www.pnas.org/cgi/doi/10.1073/
pnas.1417000112.

Chapter 22

The four theories for the origins of viruses: Fisher S., Are RNA
Viruses Vestiges of an RNA World? *J. Gen. Philos. Sci.*, 2010;
41: 121-141; Forterre P., The origin of viruses and their possible
roles in major evolutionary transitions. *Virus Research*, 2006;
117: 5-16; Bremerman H.J., Parasites at the Origin of Life.
J.Math.Biol., 1983; **16**: 165-180; Koonin E.V., Senkevich T.G.
and Dolja V.V., The ancient Virus World and the evolution of
cells. *Biology Direct*, 2006. doi:10.1186/1745-6150-1-29;
Villarreal L.P., 2005, *Viruses and the Evolution of Life*.

Life beginning as prebiotic self-replicators: Lazcano A. and Miller
S.L., The Origin, Early Evolution of Life: Prebiotic Chemistry,
and the Pre-RNA World, and Time. *Cell*, 1996; **85**: 793-798;
Cronin L., Evans A.C. and Winkler D.A., eds. 2017. From
prebiotic chemistry to molecular evolution. www.belstein–
journals/bjoc/70.

Self-replicators being parasitised by other self-replicators: Eigen
M., Self-organization of matter and the evolution of biological
macro molecules. *Naturwissenschaften*, 1971; **58**(10): 465-523.

HIV as a quasispecies: Nowak M.A., What is a Quasispecies?
TREE, 1992; 7(4): 118-121.

The RNA World: Gilbert W., The RNA world. *Nature*, 1986;
319: 618; see also Rich A., On the problems of evolution and
biochemical information transfer. *Horizons in Biochemistry*, 1962.

Kasha M. and Pullman B., eds. Academic Press, New York, pp. 103-106.

Parasitic elements arising in self-replicator experiments: Bremerman H.J., Parasites at the Origin of Life. *J.Math.Biol.*, 1983; **16**: 165-180; Colizzi E.S. and Hogeweg P., Parasites Sustain and Enhance RNA-Like Replicators through Spatial Self-Organisation. *PLOS Computational Biology*, 2016; doi:10.1371/journal.pcbi. 1004902.

Quasispecies gives individual members advantages in survival: De La Torre J.C. and Holland John J., RNA Virus Quasispecies Populations Can Suppress Vastly Superior Mutant Progeny. *J. Virol.*, 1990; **64**(12): 6278-6281.

Key viral genes absent from cellular life: Prangishvili D. and Garrett R.A., Exceptionally diverse morphotypes and genomes of crenarchaeal hyperthermophilic viruses. *Biochem. Soc. Trans.*, 2004; **32**(2): 204-208; see also Koonin, Senkevich and Dolja 2006, above.

RNA viruses originating in the RNA World: Forterre P., The origin of viruses and their possible roles in major evolutionary transitions. *Virus Research*, 2006; **117**: 5-16; see also Koonin, Senkewich and Doljva 2006.

The symbiotic virosphere: Villarreal L.P., Force for ancient and recent life: viral and stem-loop RNA consortia promote life. *Ann. New York Acad. Sci.*, 2014; **1341**: 25-34; Villarreal L.P. and Ryan F., published in the *Handbook of Astrobiology*, ed. Vera M. Kolb. CRC Press, Boca Raton Florida, 2018.

Viruses in the deep-sea hydrothermal vents: Prangishvili D. and Garrett R.A., see above.

The transfer of genetic information is far commoner from virus to host rather than from host to virus: Villarreal L.P. 2005, see

books; Filée J., Forterre P. and Laurent J., The role played by viruses in the evolution of their hosts: a view based on informational protein phylogenies. *Research in Microbiol.*, 2003; 154: 237-243; Claverie J-M, Viruses take center stage in cellular evolution. *Genome Biol.*, 2006; 7: 110. doi: 10.1186/gb-2006-7-6-110.
The addition module concept of self: Villarreal L.P. 2005, in books; Villarreal L.P. 2014.

Chapter 23

Woese's iconoclastic first paper on domains: Woese C.R. and Fox G.E., Phylogenetic structure of the prokaryotic domain: the primary kingdoms. *Proc. Natl Acad. Sci. USA*, 1977; **74**: 5088-5090.

A further elucidation of the three domains: Woese C.R., Kandler O. and Wheelis M.L., Towards a natural system of organisms: Proposal for the domains, Archaea, Bacteria and Eucarya. *Proc. Natl Acad. Sci. USA*, 1990; **87**: 4576-4579.

Are viruses alive?: See Villarreal L.P. and Ryan F., 2018, published in the *Handbook of Astrobiology*, ed. Vera M. Kolb. CRC Press, Boca Raton Florida, 2018. See also, Koonin E.V. and Dolja V., A virocentric perspective on the evolution of life. *Curr. Opin. Virol.*, 2013; **3**(5): 546-557. Villarreal L.P., Force for ancient and recent life: viral and stem-loop RNA consortia promote life. *Ann. N.Y. Acad. Sci.*, 2014; **1341**: 25-34.

More about 'extremophiles': Lindgren A.R., Buckley B.A., Eppley S.M., et al., Life on the Edge – the Biology of Organisms Inhabiting Extreme Environments: An Introduction to the Symposium. *Integrative and Comparative Biology*, 2016; **56**(4): 493-499. See also Rampelotto P.H., Extremophiles and Extreme Environments. *Life*, 2013; **3**: 482-485.

Overview of Archaea and their viruses: Snyder J.C., Bolduc B. and Young M.J., 40 years of archaeal virology: Expanding viral diversity. *Virology*, 2015; **479-480**: 369-378. Prangishvili D., Forterre P. and Garrett R.A., Viruses of the Archaea: a unifying view. *Nature Rev.*, 2006; **4**: 837-848.

The central role that viruses play in causing disease, controlling microbial community composition and structure, and driving evolution: Bolduc B., Shaunghessy D.P., Wolf Y.I., et al., Identification of novel positive-strand RNA viruses by metagenomic analysis of archaea-dominated Yellowstone hot springs. *J. Virol.*, 2012; **86**: 5562-5573.

Viruses and cells ever entwined: Durzyńska J. and Goździcka-Józefiak A., Viruses and cells intertwined since the dawn of evolution. *Virol. J.*, 2015; **12**: 169. doi 10.1186/s12985-015-0400-7.

Index

Index

Index

Index

Index

Index